# Material ConneXion

## THE GLOBAL RESOURCE OF NEW AND INNOVATIVE MATERIALS
## FOR ARCHITECTS, ARTISTS AND DESIGNERS

# Material ConneXion

THE GLOBAL RESOURCE OF NEW AND INNOVATIVE MATERIALS
FOR ARCHITECTS, ARTISTS AND DESIGNERS

**GEORGE M. BEYLERIAN**
**ANDREW DENT**

EDITED BY ANITA MORYADAS

Thames & Hudson

First published in the United Kingdom
in 2005 by
Thames & Hudson Ltd
181A High Holborn
London WC1V 7QX

www.thamesandhudson.com

© 2005 Material ConneXion

British Library
Cataloguing-in-Publication Data
A catalogue record for this book is
available from the British Library

ISBN-13:  978-0-500-51244-9
ISBN-10:  0-500-51244-2

Design Alexander Boxill
Printed and bound in China

# LIVING IN A
# MATERIAL WORLD

Materials surround us.

Everything we hold, everything we see, takes shape thanks to materials. Materials will always be around us. We experience how they are evolving every day.

Look around and you will see the world of materials we frequently take for granted: wood, polymer, glass, ceramic and so on. How often do we notice the variety of textures, finishes and colors? Since materials are everywhere, we rarely pause to truly think about them. And yet they help to sustain our life and increase productivity, and they serve a variety of critical human needs.

We have learned how a great design becomes successful if the right material is selected.

I strongly believe that material equals design.

Material ConneXion was founded to provide innovative materials to specifiers across many diverse industries, from automotive to apparel. This library, with several thousand samples, has become a forum from which to launch new materials and accelerate new business opportunities.

Material ConneXion has created the opportunity for anyone involved in a creative endeavor to learn and find materials and processes to be applied to their specific field. For everyone from students to executives, the information is provided in a simple and accessible manner, thus creating an international hub where developers and specifiers can meet.

Nowadays the creation of products demands not just affordable prices but quality. In this arena, materials play a very important role.

Materials are not trendy; they are a necessity for the realization of creativity. It is up to everyone to create a better environment by choosing the right material for the right product.

We will always live in a material world.

**MICHELE CANIATO**

# THE POWER OF MATERIALS

The refinement and transformation of raw materials are essential signs of civilization. The earliest tools were derived from materials that were found, developed and fine-tuned over time. Flint, for example, was found to be useful in order to improve the quality of human life by both making the ignition of fire comparatively easy and making the act of cutting possible. Materials – how they were formed and shaped, what was made of and with them – have been one of the essential constants throughout human history.

As our forebears found and shaped materials into tools and other necessary items, they developed 'professions' which were defined by those indivduals who knew what materials to use, and how, in order to make objects of utility or beauty.

During the nineteenth century, for instance, craftsmen and -women in France produced baskets made of woven steel wire (page 9). They wove the wire over a 'form', thus producing the shape of the finished product. Contemplating such objects makes us wonder how materials and those who shape them (in one form or other) form civilizations. People who do things with materials not only add to our culture but also, in the process, are able to make a living.

Of course ideas are always changing, and innovations continually transform our lives and cultures. In the West today there are a very few people who are able to make their living just from working with

wire. The same thing is true of basket-weaving: though still prevalent in craft circles, the idea of a basket repairperson has disappeared. In fact numerous materials-based professions simply vanished in the last century. Raw materials and the ways they are processed formed the core of many of those lost (but not forgotten) modes of employment.

Such transformations are ongoing and can be seen in the continuing changes to the world's economy and the evolution of its labor force. China, awakening from strict Communism, has become one of the largest consumer-goods suppliers in the world. While this massive source of labor supplies products at competitive prices, the United States has retained its leading position in the area of high technology. Even if China is building up its own technological resources, the power of American spending on research and development is vastly greater.

For many years I have been observing the flow of interest in, and experimentation on, materials by design professionals and artists. Although not always knowledgeable about the world of science, artists are the most prone to experimenting with new materials and forms, seeking new media to achieve their creative goals. Some are obsessed for extended periods of time, like Picasso with his ceramics and pottery. Others are more aggressive in the use of materials for their 'inventions'.

Let us probe the works of various creative people who have used materials intelligently, provocatively and innovatively. I have personally observed a certain tendency, maybe a natural evolution, in where and how things happen. Just like prehistoric humans who found flint and shaped tools out of it, our cultural evolution has brought us to a world in which material innovation is the major symbol of progress. The 'invention' or selection of a material has much to do with 'winning' at the end of the day in the world of commerce, with distinguishing the ingenious individual from the crowd in art, fashion, buildings and products. Some of these inventions are best analyzed as the wonders of the time in which we live.

Whether it's economic or cultural pressures, or the professions that produce the talent of innovation and the innovators themselves, one needs to observe how things evolve. My own observations have led me to think that artists are usually the first ones to delve into material

experimentation. There are artists of
international fame who have made
incredible works with exciting media, as
well as artists maybe of lesser fame who
have made great creative contributions.
There are of course craftsmen, who used
not to be considered fine artists, a highly
debatable issue best discussed
elsewhere. Then there are product
designers, who act like artists, creating
masterpieces. Benvenuto Cellini was a
Renaissance silversmith, but he so
excelled at his craft that the cultural
world would certainly call him an artist,
not a silversmith or a metalworker.

This essay is not about how to label
people, and not about the full list of
unbelievably talented creators. It is
written to help us understand how craft,
workmanship and material – all combined
– each have an incredible impact on the
final 'label' one assigns to a work or
product.

**WIRE BASKET**
A turn-of-the-century
wire basket made by
hand in France.

It is appropriate to intervene here with an idea that has puzzled many people. It's the old 'fish or fowl' syndrome that arises when trying to define the profile of a multi-talented person with a high level of productivity across several fields. How do we categorize such individuals? Some are trained in many fields; some are not formally trained in any but excel in several (call them 'talented'). Other such 'talented' individuals become bored with their early work and move along to other fields in which they excel (trained or not).

Let's take Michael Graves as an example. My first exposure to Graves was a temporary 'installation' he had done for Sunar at the Pacific Design Center in Los Angeles. It was a most ingenious solution to a temporary need. A brilliant solution. The work was unlike anything else he'd ever done. When I then went to see his new work at the Merchandise Mart in Chicago, I found, much to my surprise, something completely different from his Los Angeles masterpiece. Here was an incredibly visual spatial experience blasting with colors, a maze-like tour of a blend of ancient Egyptian and Assyrian architecture, all combined with the utmost dexterity. All this effort and bravura was the setting for 'contract furniture'. This was a cultural 'trip' that nobody would ever forget. Was it architecture or some other profession for which no description has yet been coined? Shortly thereafter, I discovered Graves' show of beautiful watercolors at the Max Protech Gallery in New York: a reverie of Tuscan and Roman pastoral scenery with

lots of delightful sketches of classic architectural elements.

Is Graves an architect, a painter, a product designer? We are pleasantly baffled. His enormous current success is the myriad of products he has designed for Target. Many members of a mass audience will recognize his name for work he was not formally trained to make. His oeuvre defies simple categorization.

Our second example is the indomitable Gaetano Pesce (his name says he's a FISH – or is he a FOWL?). Pesce has an incredible knack for material experimentation, a form of play with things others haven't tried, or thought to try, before him. His use of materials, sense of symbolism, manipulation of color and reverence for historical references all add up to his genius.

Pesce's original training, as with many other material people of the day, was in architecture. So many Italian architects were trained without the chance to put their skills to work toward buildings, especially post-World War II. Their key to success has often been through industrial design to meet emerging social needs. Hence the strong sense for useful objects of sensual beauty and furnishings that have caused Italy's second Renaissance.

Pesce is an idea man: he knows he's great. Anybody who can create an armchair out of spaghetti must have considerable talent. My first exposure to his early work was possibly the Golgotha

1 **FLEXFORM SPECIAL FINISH**
This boom box, created by Philips for Target, has a pearlescent finish.

2 *UNTITLED*, 1968
Original idea and lustrous finish developed by Craig Kauffman.

3 **I FELTRI**, 1987
An armchair made of felt, designed by Gaetano Pesce for Cassina, that is an industrial product yet has a hand-made look.

2

chair made of resin-dipped fabric. In contrast to his other 'crafted-looking' pieces was the highly industrialized Blow Up series – the B & B-produced curvaceous Bambole armchair and hassock in a ball shape (now enjoying a re-edition). Further along in his exploits was the Feltri chair (page 11/3) for Cassina, to be produced according to the principle of utilizing inexpensive manpower in Third-World countries to turn out furniture at low prices. Great concept, but that's not how the chair ended up being so successful. Its success was due to its superb execution in Italy using felt as its main material and its being produced quasi-industrially to look like villagers had crafted it.

The examples of Graves and Pesce lead us to yet another dilemma of nomenclature. Graves' products for Target may have become staples for a mass audience, but does that audience recognize him? Is that important (for Graves it might be)? But is it important culturally, and for the sake of history? Then we turn to Pesce, who thinks he is designing for the masses (Fish Design), as well as creating a new collection of major pieces of furniture, mostly for collectors. The Fish designs are visually very exciting; whether they are for the masses or not, their cultural impact is substantial. The critical factor is the genius Pesce brings to an object through his choice of materials, the maneuvering and shaping of them, the low cost/high effect of the final product.
Which of these two talents is making an

impact on our evolution in terms of design, style, symbolism and innovation? The use of the more contemporary materials in Pesce's work has a great influence on our society, as it has an impact on the future of global cultural values. While materials have clearly made an impact on Pesce, Graves' work for the mass market is an incredible mediator of contemporary taste.

But what do we call these two talented people as we approach other examples of individuals in various fields of specialization? Both are architect/designers, of course. But for the purpose of this book, I wish to be descriptive of all their proven accomplishments. Though I would not call Graves an industrial designer, he is producing design for products. He has used materials intelligently though not provocatively. Not all mass-produced products are products of 'industrial design', since they don't involve complex problem-solving and intense material acuity. The 'industrial' idea has taken on a new meaning: the involvement of many product-development techniques.

For Pesce, on the other hand, materials

play a large role in almost every product and project he has designed. In my book he is a material architect and designer. Anybody worthy of the label 'material innovator' should have the word *material* in front of his or her *métier*. Since materials provide a basic way to see the world around us, we are really looking at how the world is evolving.

1 *WALTZ OF THE POLYPEPTIDES*, 2003
Mara G. Haseltine uses contemporary commercial finishes when creating her outdoor sculptures.

2 *RED TREE*, 1998
Charles Worthen has used household silicone caulking as the sole medium for a large collection of sculptures.

3 **GEL MASK**, 2001
Gaetano Pesce designed this mask as an invitation to an exhibition. Various resin colours were poured simultaneously into a hand-made mold, giving each invitation a unique, individual look.

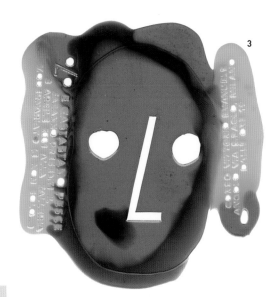

Of ARTISTS working today, an excellent example of someone who has been an innovator through materials and process is **CRAIG KAUFFMAN**. I have personally lived with, and love, one of his works created over thirty-five years ago. This beautiful 'bubble' (page 11/2) is iridescent; visitors often mistake it for a lamp. The process Kauffman created for this finish has been a mystery – a specialty of his own. Not for long! We now have in our library at Material ConneXion a commercially available finish that imitates the Craig Kauffman look. An excellent example of how art creates, through the years, a product for the masses. I say 'masses' because this new finish, from Flexform, has been applied to a radio/music box created by Philips for Target, who selected it in consultation with Material ConneXion (page 10/1).

Another artist who works with a similar finish, but not invented by her this time around, is **MARA G. HASELTINE** (page 13/1). **CHARLES WORTHEN**, on the other hand, is passionate about silicone (page 13/2). In this case, the material best explains the artist's message – what he is trying to communicate through material and form.

**GAETANO PESCE**, in addition to his achievements as a material architect, has made many pieces out of urethane, rags, glass, metal and so on. The mask illustrated here (page 13/3) conveys a typical message: a new medium calling for attention, provocative enough to be called art – but it's actually a promotional piece.

**GREGORY M. BEYLERIAN**, a self-made artist in multimedia (but trained as an industrial designer), plays with materials, photography, words and found objects. In the work shown here (page 15), he has made a 'living object' – actually a vintage Volvo covered with a hundred plush elephants which had been discarded in front of a toy factory. The Volvo's dashboard is covered with ethnic paraphernalia that expresses Beylerian's Buddhist beliefs, as well as other related religious and romantic items.

**KARDASH ONNIG** is a sculptor of grand-scale work. The toys he has created seem primitive but are actually extremely sophisticated in their simplicity. In his *Mountain Goat* (page 17/5) he used goatskin over the wooden toy itself. Not an unusual happenstance, but it's the reality of the application that makes this object so unique. A special sensibility that only an artist could possess, made manifest in a mix of material, creativity and sensitivity.

**AIDA GOGANIAN** is a California artist who has worked intensely with discarded materials, which she transforms through various manipulations into a more 'noble' state, one of which is braids: woven 'textiles' mixed with various fibers. This process of recycling to produce new art pieces may not be unique, but when done in a consistent manner, it becomes an art form all in itself. The Geisha doll (page 17/4) illustrates Goganian's use of a cigar box, coins and various discarded items, clearly showing that new things can come out of old paraphernalia – a form of ecological thinking.

**ROBERT EBENDORF** is a remarkable artist specializing mostly in jewelry who excels in the exploration of design through the use of recycled materials. Shown here are a necklace that has a pendant made of crab claws, and his famous 'Off the Street, From the Beach' necklace (page 17/2+3). The idea of incorporating mundane and junk pieces into works of fine art is truly remarkable – it almost drives us to wonder if technology is really necessary.

*MEDZMOBILE*, 1999–
Gregory M. Beylerian
is continually adding
new elements and
making changes to
this living work.

Architecture represents a vast opportunity for the application of new materials. Many <u>ARCHITECTS</u> are fascinated by the idea of using new technologies and materials in construction. However, few end up doing so for a variety of reasons such as difficulty with building codes, resistance from construction crews, and union demands which lengthen the building process. Many situations require short-cut solutions, so architects revert to standard methods and methods to get projects going. Design innovation is surely a useful tool, but when technology is incorporated in the form of new materials and processes, the results are always noticeable. We call architects who go out of their way to do this 'material architects'. This means those who brave the difficulties of innovation, taking major steps forward. Many of these material architects have earned kudos for their bravery. Among them are Frank O. Gehry, Rem Koolhaas, Jean Nouvel, Sir Norman Foster, Toyo Ito, the firm of Herzog and de Meuron, Shigeru Ban, Steven Holl, Toshiko Mori, Peter Zumthor, Renzo Piano, Nicholas Grimshaw and quite a few more. If we look at those individuals or firms that have had the courage to make the special effort to bring innovation to the state-of-the-art level, we see that pushing the edge of the envelope in major construction projects is no easy task.

In a 'non-material' way, like the artists already mentioned, architecture has its own angels of reverse technology. The late <u>**SAMUEL MOCKBEE**</u>, a major figure in the world of 'natural' architecture, produced, through his <u>**RURAL STUDIO**</u> at Auburn University in Alabama, numerous works revealing his dedication to the New Architecture of Social Justice. Illustrated here is his Cardboard Pod (page 16/1), made out of bales of recycled, wax-impregnated cardboard.

In a similar way, but using mostly abandoned or recycled shipping containers and other discarded materials, <u>**LOT-EK**</u> (Ada Tolla and Giuseppe Lignano) have made waves on the architecture scene (page 19/4). Their bold statements have included twisting the idea of material-as-innovation, working with used and battered pieces of yesteryear. It should be noted, however, that in most cases these inventive applications work best in the context of conceptual and creative endeavors, and must be considered as a source of inspiration. They don't signify that mass production will ever depend on existing recyclable materials.

Of particular interest is the work of <u>**KIERAN TIMBERLAKE ASSOCIATES**</u>, who experiment a great deal with new materials. True to tradition, and in the spirit of Louis Kahn's soliloquy with bricks, they have created a feast of brick patching for Little Hall at Princeton University in New Jersey (page 18/1). It is not often that one can find a single material juxtaposed in so many ways that each iteration could indeed become a work of art.

Inflatable materials and ethyl tetrafluoroethylene (ETFE) have been used experimentally and practically in Europe and in Japan. <u>**FESTO AG & CO.**</u> have promoted their Airtecture structure (page 19/2), an air-inflated series of chambers making up an exhibition hall. This concept exploits new techniques and materials. Also of interest is the Airquarium, which is 105 ft/32 m in diameter and 26 ft/8 m high (page 19/3). An even more sophisticated use of ETFE is the Rocket Tower by Nicholas Grimshaw and Partners, located at the Leicester Space Centre in the UK; it uses double-sided inflatable panels of ETFE.

Shigeru Ban is an exceptional architect who uses the simplest and most natural materials in his architecture. Paper, wood and bamboo are major ingredients. These have been employed most successfully in a range of applications, from his paper house (using paper tubes) to emergency shelters and exhibition spaces; the latter included the Japanese pavilion at Expo 2000 in Hanover and the outdoor space at the Museum of Modern Art in New York before it closed for its major renovation. The incredible audacity involved in taking paper tubes, or even bamboo, and creating serious architectural spaces, whether permanent or temporary, is a major stretch of the imagination in the use of materials.

2 **OFF THE STREET, FROM THE BEACH, C. 1992**
Robert Ebendorf's necklace is made from *objets trouvés* found while out walking.

3 **CRAB-CLAW NECKLACE, C. 1996**
Ebendorf takes an ecological approach, wasting nothing.

4 **GEISHA DOLL, 2003**
Aida Goganian uses household refuse and adds the occasional piece of paraphernalia to create objects charged with memory.

5 **MOUNTAIN GOAT, 1983**
Kardash Onnig carved a goat as a toy and then covered it with real goatskin from an animal his son and friends had played with.

1

There is a strong sense of creativity, inspiration and daring in the spirit of some FASHION DESIGNERS. My theory is that they run very close to the fine artists in bringing their material aspirations, and voyeurism, into the field of clothing. Basic designers would generally be interested in using good fabrics, yarns and so on as provided by mills and the usual fashion-world suppliers. However, those designers who have been singularly creative, such as Hussein Chalayan, Jean-Paul Gaultier, Dries Van Noten and the master of them all, ISSEY MIYAKE, have opened up the horizon by exploring material juxtapositions. They have also given the public a sense of security when learning about the creativity that new materials and fibers can bring. These creations range from the ridiculous to the sublime and sometimes reflect the zeitgeist. Inflatable garments were an expression of the 1960s – for instance, the daring 'discs' of Paco Rabanne. While the '60s may have been more radical in experimenting with new materials, contemporary designers play with the poetic nuances as well as the technology that comes with garments, such as the famous pleats of Miyake (page 20). Innovation goes hand in hand with the technology of the material world in works like Miyake's inventions.

In the last four years, the industrial-design profession has clearly advanced to pole position. The reason for this is clear. As industrial design has become much-coveted territory in both education and practice, INDUSTRIAL DESIGNERS now need to distinguish their work from that of others by thinking big in the areas of design and development. It is said that a product is the symbol of a company's capabilities. Designers nowadays are faced with having to show their smartness, alacrity and emotional intelligence by expressing their knowledge of the technological features and performance of various new materials. This is the way to win over the competition: performance, performance, performance. Other choices available to contemporary designers are innovation, ecological features, new processes, transparency, conductivity, originality and so on.

Appropriateness of materials is a major factor in this area. So many of our daily necessities are a product of industrial design; hence this is now a very special category in which material excellence is absolutely a major factor. Here again, there are designers who excel in their ability to combine sensitive material choices and design. We call this talented group 'material designers', which, to us, means that they are designers with a cultivated eye and a good understanding of the availability, proper selection and application of materials. Some bring poetic license to their designs while others use materials in a hard-core way,

as purely functional. For instance, REIKO SUDO, known for her work at Nuno, is mainly a textile guru (page 21). To quote Ryu Niimi, curator of a 2003 traveling exhibition on Japanese style:

'She weaves in a seemingly inexhaustible range of materials, such as copper wire, paper, rubber, feathers and iron rust; she makes things that are transparent, fraying, folded, closing and creasing. There are supreme and gorgeous metamorphoses in her work. We see a fusion of local materials and cutting-edge technology, even a renovation in local technological skills through the innovative use of novel materials. It is the destiny of textile art never to allow material to be separated from technology. They exist before the eye together, allowing an excitement that seems improvised; for all that we know there is an extremely lengthy process behind the planning, preparation and trial manufacture of these items, and of collaboration with factories and sources of materials.'

**1 LITTLE HALL, PRINCETON UNIVERSITY**
Kieran Timberlake's ingenious use of brick in the renovation of Little Hall created privacy while retaining the original archways.

**2+3 AIRTECTURE + AIRQUARIUM**
ETFE is used to create semipermanent architectural envelopes.

**4 MOBILE DWELLING UNIT**
Lot-Ek has pioneered the re-use of shipping containers to give them a new function, in this case as a modular home.

The above description is most appropriate for emulation by every industrial designer. It explains the exploration of materials, and their sensitive role in the integration of material and design. It is the utmost clarity of the material(s) *vis-à-vis* the final project that makes a most successful product. Totally transformed, and totally beyond mere 'textiles', is Sudo's Scrapyard, 2003 (page 21).

Industrial designers have the challenge to meet clients' rigorous needs. It's about performance, originality, price and quality all integrated into one hard-core assignment.

Who would believe that someone with a phobia for science, technology and chemistry would delve into a venture based on those very skills and knowledge? In 1997 I was 'driven' to create a company that would connect the world of science with the diverse audiences that seem to have a need for scientific information. Most of those people would fall into the category in which I myself belonged. I am using the past tense not because my knowledge has increased, but because I have conquered part of the anxiety caused by the vast amount of knowledge out there that is still totally foreign to me. The truth is that

I can now tackle the world of technology because I am able to recognize materials through the lens of a more familiar vocabulary. This lens – and my inspiration in creating Material ConneXion – is a precious tool that uses simple vocabulary to explain the phenomena of the world of science.

**GEORGE M. BEYLERIAN**

**SCRAPYARD, 2003**
Textile magician
Reiko Sudo invents
new techniques and
processes for
constructing fabric,
with innovative
results.

INTRODUCTION
USING THIS BOOK

21

# USING THIS BOOK

The seven materials sections of this book have been organized to follow the model of the Material ConneXion Library. The hyphenated Material ConneXion Index number for each material is included at the end of the relevant caption. The seven categories allow all of the materials to be grouped according to chemical composition. The reason for this is to avoid defining materials in terms of current use, instead giving information about their base compositions and thereby providing a greater understanding of limitations and potential. Composite materials such as glass fiber-reinforced plastics or terrazzo are categorized by the material that comprises 51 per cent or more of the total. It should be noted that this form of categorization is just one of

many ways in which materials and technologies may be defined. But it has helped a new generation of designers to think 'materially' when creating solutions to today's problems.

To maintain objectivity, all materials in the Material ConneXion Library have been voted in by a panel of senior creative professionals who have judged the innovative qualities of each one individually. The panel, which sits monthly, includes both permanent and invited jurors. Judging innovation is neither a simple task nor an entirely objective one. The intended result is the acceptance of materials or technologies that have demonstrated use of a new material; offer an improvement on an existing process; create a solution through the use of less material or through a more sustainable methodology; or simply offer a solution that already exists in another industry. It is this last criterion that has provoked the most intriguing solutions to material issues, and such cross-fertilization of ideas continues to be a integral part of the Material ConneXion ethos.

The process of selection is not exhaustive. There are thousands of new materials produced each year, and to catalogue them all would diminish the selections that are made. Most are chosen with an eye to their intended audience, which includes artists; fashion designers; interior, product and graphic designers; sneaker, automotive and cosmetics-packaging designers; and architects.

Above all, the intention has been to create a library of materials and technologies that informs, inspires and allows easy access to the most innovative and newest materials from around the world.

**ANDREW DENT**

01
CARBON-BASED

DIMENSION
POLYANT
SAILCLOTH
A laminate structure
that incorporates
carbon fibers in both
the warp and the
weft to create high
tensile-strength
sailcloth. The high
strength-to-weight
ratio and variable
density of the weave
make it an attractive
alternative to glass
fiber. Although
primarily used for
high-performance
sailing applications,
it could be used in
buildings as well.

The materials sourced for this category are dominated by one type: carbon fiber. From its early beginnings as a strengthening fiber in the late 1960s, carbon fiber maintained an aura of high performance hallmarked by the distinctive deep black color of the filaments. These filaments are created by heating pitch (tar) or a polymer called polyacrylonitrile to form ribbons of graphite, which pack together to form the fiber. Weaving, braiding, knitting or winding the fibers can create a range of different forms, including textile-like mats, cylinders, ropes, bricks and even more complex forms such as cones. The fibers are made rigid by coating or immersing them in an epoxy, polyester or polyurethane resin, a glue-like substance which hardens and maintains them in their intended form. The different resins offer varying properties such as flexibility, toughness or rigidity, but all function as binders, allowing the carbon filaments to act as the stiffening and high tensile-strength elements.

With high performance comes corresponding high costs. As a result, carbon fiber is still the preserve of aerospace design, racing yachts and top-of-the-line sports equipment. Of course, ever keen to utilize the newest and most innovative materials, furniture designers have created super-strong chairs from woven carbon fibers. Developments in processing have also reduced the cost of certain carbon-fiber products. Processes such as 'liquid-

infusion technology' (page 29/5) enable complex solid forms to be constructed with an outer carbon-fiber skin and a rigid polyurethane-foam core, dramatically reducing the cost of manufacture for furniture such as stacking chairs and sports equipment such as hockey sticks.

The relative youth of this material compared to glass, metal and natural materials has meant that there are still many areas of application to be investigated. The ability to produce unlimited amounts of continuous-length filaments has led to the consideration of carbon-fiber objects of greater magnitude than are being produced currently. One interesting example might be the use of millions of continuous strands wrapped in a spiral around a frame to create cylindrical buildings. The fibers would be epoxy-coated in situ as they were laid over a frame, thus creating a building in which all the stresses were concentrated in the outer skin rather than in vertical struts.

In addition to its exceptionally high tensile strength-to-weight ratio, carbon also

possesses high thermal and electrical conductivity; this has been used to great effect with pressed graphite foams (page 26/2). These 'thermal-management foams' can be machined to tight tolerances and are both light and hard. (The resistance of graphite to elevated temperatures makes it an ideal high-temperature insulation material, as is found with some reticulated-cell foams for pipe lagging [page 26/1]).

The next generation of carbon-based materials is likely to incorporate carbon nano-tubes. These cylindrical sleeves of graphite, also known as buckytubes and discovered in 1991, are in fact elongated buckyballs (named after Buckminster Fuller's geodesic spheres), approximately 1 nm in diameter but found up to 0.08 in./2 mm in length. The outstanding properties of these tiny tubes are still being fully realized, but their exceptional electrical and thermal conductivity, strength, stiffness and toughness have led to research into their use as spun fibers and also as fiber-reinforced forms that offer great improvements over current carbon-fiber composite materials. Carbon still has a lot more to offer in the field of high-performance innovation.

**TALON**
CLASSIC CHAIR
This chair costs a fraction of the price of any other carbon-fiber chair due to the unique construction of the legs and back. The carbon fiber is actually a sleeve into which polyurethane foam is injected during the manufacturing process. This makes possible far more complex shapes than can be achieved with carbon fiber alone.

**01**
**CARBON-BASED**
GRAPHITE
INTUMESCENT

**1 CARBON FOAM**
Open-pore foam composed of glass-like carbon. The material is conductive, has high-temperature strength and low-bulk thermal conductivity, and is very chemically inert over a wide temperature range.  129-02

**2 CARBON FOAM**
Lightweight graphite foam for thermal-management applications. The microcellular structure can be machined to tight tolerances for use in electronics.  4714-01

**3 CARBON INTUMESCENT**
Chemically inert, highly stable, expandable door seal composed of graphite in a polymer binder. When the temperature rises to 400°F/204°C, the seal expands multi-directionally to fill the gap around the perimeter of the door.  3728-01

**4 CARBON-FIBER COMPOSITE**
Custom flexible thermoplastic sheet materials that are durable and impact-resistant with damping and shock-attenuation properties. They are composed of a proprietary poly(methyl methacrylate) resin system reinforced with fibers, such as carbon, by 65 to 75 per cent of volume. 112-02

**5 CARBON FIBER**
Carbon-fiber filaments braided into shapes for composite reinforcement. Various shapes are possible, including sleevings, tapes, tubes and net-shaped pre-forms made rigid by using a polymer resin. 1254-01

**6 CARBON-FIBER COMPOSITE**
Woven carbon fiber-reinforced thermoplastic for high-performance applications. The rigid composite sheets are durable, impact-resistant and have good shock-absorbing properties. 112-01

01
**CARBON-BASED**
WOVEN CARBON FIBERS

1 **CARBON-FIBER REINFORCEMENT**
A range of engineered, multi-axial reinforcement fabrics constructed as a matrix of stitch-bonded layers of unidirectional fibers (E-glass, carbon, aramid) in various sizes, weights and fiber orientations. The fabrics can be draped and conformed. 1277-01

2+3 **CARBON-FIBER COMPOSITES**
A range of high-performance woven Kevlar® and carbon-fiber fabrics and tapes coated with a clear polyurethane laminate. 4569-01

4 **CARBON-FIBER REINFORCEMENT**
Bi-directional fabric tapes woven of fiberglass, carbon/graphite, aramid (Kevlar®), ceramic and quartz, in plain, basket or leno weaves, depending on the application. For use as reinforcement and as electrical and thermal insulation. 2251-01

1

2

3

4

5

6

**6 CARBON-FIBER
COMPOSITE**
Carbon fiber-and-resin-
composite system for the
repair and strengthening
of concrete structures.
This system comprises
three layers of stitched
carbon-fiber reinforcement,
a primer to prepare the
concrete, putty and a
solidifying resin.  4822-01

**5 CARBON-FIBER
COMPOSITE**
A process that creates
low-cost woven, fiber-
reinforced composites for
sporting and furniture
applications. The woven
reinforcement is inserted
as a sleeve into a mold. An
inserted bladder then
forces the sleeve to
conform to the mold's
shape.  4931-01

02
CEMENT

↙

This category covers not only cements but also some of the additives and fillers that are used to change their colors or strengthen or improve their appearance. The materials are based on Portland cement – made by heating powdered clay and limestone together – which is then pulverized and mixed with gypsum. The primary constituent of concrete, Portland cement is also used in mortars and grouts.

A spectrum of pigments is available for coloring concrete (page 34/3) that is light-fast, alkali- and weather-resistant, and permanent. The pigments are metal or mineral oxides. They are supplied as free-flowing granules, liquids formulated with high pigment content or concentrated powder pigments. The use of iron oxide in sufficient quantities as a pigment makes it possible to create the appearance of rusting on the surface of cement.

The development of cements that are not corrosive to glass has led to a range of cement-based construction materials, terrazzo-like tiles and cement blocks that incorporate recycled glass chips for strengthening, filling and decorative effect. The use of cement as a binder for non-combustible paneling has led to corrugated, colored (page 34/3) and lightweight sheets that can be used as wall and floor surfaces. These fiber-reinforced cement sheets, composed of a compound of cement, silicon and calcium strengthened with cellulose fibers and resins, are asbestos-free. The sheets are noncombustible, rot-, fungus- and vermin-proof, and resistant to freezing. Quick-curing veneers have also been produced using this type of material, thus offering thin cement surfaces that may be applied to cardboard honeycomb or MDF panels to create lightweight paneling for walls or countertops.

**CEMENT TILE**
Different looks can easily
be created in cement by
altering the aggregate
pigmentation and surface
texture. Cement improves
its strengh in compression.

**1 COLORED CONCRETE**
Lightweight, iron oxide-pigmented concrete. It is two-thirds of the weight of standard concrete. Its acrylic sealant (available in either matte or gloss) can withstand temperatures of up to 330°F/166°C.  4605-01

**2 SELF-LEVELING CEMENT**
A self-drying, cement-based finishing underlayment that provides an even, permanent finish by smoothing ridges and filling cracks, gouges and joints. It is appropriate for all types of interior concrete, wood, cementitious terrazzo, and ceramic and quarry tile. 1101-03

**3 COLORED CONCRETE**
Smooth-surfaced concrete infused with color. A proprietary process is used to mold this hand-made concrete and give it a veined or troweled appearance. Then the surface is ground with water to expose the natural aggregate. 4038-01

**4 PRECAST CONCRETE TILE**
A durable, precast polymer-concrete tile with a low resin content. Designed as a detectable warning surface for the blind, visually impaired and handicapped, it is manufactured in warning yellow or almost any other color.  39-01

1

2

4

3

5

**7 SELF-LEVELING CEMENT**

A self-drying, self-leveling, Portland cement-based topping for fast-track resurfacing, smoothing or leveling of indoor concrete. It hardens fast and does not shrink, crack or spall. It can be installed over concrete subfloors. 1101-01

**8 REFRACTORY TILE**

Moldable refractory insulation for high thermal-shock applications. Ninety per cent alumina/10 per cent silica blankets composed of long, high-strength ceramic fibers in an inorganic binder/rigidizer. The blankets are 100 per cent inorganic and non-combustible. 3695-02

**5 SELF-LEVELING CEMENT**

An underlayment that levels and smoothes concrete and other indoor subfloors. When mixed with water, this specially formulated Portland-cement blend becomes a liquid compound that seeks its own level and produces a smooth, flat surface. 1101-02

**6 PRECAST CONCRETE**

A precast, lightweight concrete reinforced with recycled fibers and composed of natural and recycled or recovered materials from industrial and post-consumer goods. Applications include floor tiles, countertops, tabletops, fountains, sinks and bathtubs. 63-01

7

8

35

**1 SOLID SURFACING**
A tough, durable, slate-like material with either a smooth or a rough surface, made of a mixture of Portland cement, silica sand, Wollastonite – a fibrous mineral – and natural and man-made fillers, all consolidated under high pressure. 4181-01

**2 REFRACTORY TILE**
Durable, machinable, chemically inert, non-asbestos-containing cement sheets that can withstand thermal shock (within specified temperature ranges); composed of a high-alumina (aluminum oxide) content reinforced with inorganic fibers. 174-01

**3 STRUCTURAL WALLBOARD**
A structural, non-combustible sheathing composed of Portland cement, mineralized cellulose fibers from recycled materials, and mineral additives. It is weather-resistant and impervious to vermin and insects. 2144-01

**4 CERAMIC CLADDING**
A fiber-cement exterior siding that looks like wood. This durable lap and vertical siding is composed of Portland cement, ground sand, cellulose fiber, additives and water. It is formaldehyde-, fiberglass- and asbestos-free, and non-combustible. 1032-01

**5 COMPOSITE WALLBOARD**
Fiber-reinforced cement sheets for interior or exterior application. Silicon-calcium, the compound of cement, is strengthened with cellulose fibers and resin to make this non-combustible, rot-, fungus- and vermin-proof material. 4961-01

**6 CONCRETE SURFACING**
Solid surfacing from cement-based materials. This cast surface is composed of 25 per cent cement, 55 per cent aggregate, and 20 per cent fibers and fillers.  4831-01

**7 GROUT**
A polymer-modified, Portland cement-based grout that can be used with all types of tile in joint widths from 1/8 to 5/8 in./0.32 to 1.6 cm. It helps eliminate shading and powdered joints and is available in thirty-six colors.  46-02

**8 WATER-PERMEABLE CONCRETE**
A concrete-alternative, water-permeable flooring. Composed of stone aggregates and agglutinates but without sand, this exterior flooring product is, like concrete, mixed and poured.
4830-01

**9 WATERPROOFED CONCRETE**
A permanent concrete waterproofing system composed of Portland cement, very fine treated silica sand and various proprietary chemicals. It protects concrete and reinforcing steel from deterioration and oxidation.  3677-01

Ceramics can be defined very broadly as inorganic, non-metallic materials; normally crystalline, they are compounds of metal and non-metal elements. Well known for their resistance to elevated temperatures (above 6,800°F/3,500°C in some cases), these materials are also electrical and thermal insulators, have excellent chemical resistance and are relatively strong in compression. The primary drawback for almost all ceramics is their almost complete lack of ductility, which makes them brittle. Ceramics are generally prone to thermal shock and have low tensile strengths.

For many types of ceramic, aesthetics and basic functionality have been of primary importance; these have included fired clay and porcelain dinnerware, sanitaryware, artware tiling and porcelain enamels. There are also ceramics for which performance is the primary issue; applications range from the external insulating tiles on the Space Shuttle to superconductors or to the clear, unscratchable face of a Rolex® watch. It is this latter area of ceramics development that has shown the greatest innovations in recent years, with the development of such applications as flexible ceramic textiles, piezoelectrics and geopolymers.

Good thermal properties (the ability to withstand elevated temperatures), good insulation and resistance to shock have resulted in applications for the hot sections of gas turbine engines, furnaces and thermal-barrier textiles. The latter, typically made from silica ($SiO_2$), zirconia ($ZrO_2$) (page 44/7) or alumina ($Al_2O_3$) (page 44/1), offer flame-resistance as well as significant flexibility; they are used as welding curtains, as moldable furnace linings and as flame barriers in aircraft walls.

Certain ceramics, the best known of which is perovskite, exhibit the property of piezoelectricity, converting electrical energy into mechanical energy and vice versa (page 44/9). This property has found application in microphones, as ignitors in fuel lighters and welding equipment, and as sensors and actuators. Such materials are also able to store electrical energy, acting as solid-state batteries with many interesting applications (for instance, the pounding footfalls of a runner charging up a battery in a shoe that can be used to power night-lights).

Geopolymer compounds have the high-temperature resistance of ceramics and the processability of polymers. These compounds are based on alumino-silicates, with different properties of the materials dependent upon the aluminum-to-silicon ratio. The material is available in a resin form and is processed similarly to fiber-reinforced polymer composite (FRP's), pouring the geopolymer onto woven layers of ceramic or glass reinforcement. The compound is then cured at 176°F/80°C (as opposed to over 1,832°F/1,000°C for normal ceramic firing). The cured compound withstands heat of up to 2,192°F/1,200°C and may be made tougher by using various inorganic fiber fillers and particulates. Geopolymer compounds are resistant to most solvents and alkalis (though may break down in the presence of strong hydrochloric acid) and will not ignite in the presence of any flame. These materials may be applied by a range of processes including brush, pouring (casting), spray deposition and roller. Current applications for these amazing new hybrids include fire-resistant coatings for telegraph poles in areas susceptible to brush fires.

The brittleness of ceramics has precluded their use in most structural applications, as they offer the risk of catastrophic failure. However, advances in the deposition of thin layers of ceramic coatings on more ductile metals have led to the creation of scratch- and heat-resistant surfaces on less expensive metal substrates. These have been used to great effect in many extreme temperature applications such as the hot sections in jet engines, as well as to avoid excessive wear or corrosion of parts in chemical- and metal-processing plants. The coating of titanium hip- and shoulder-replacement joints with the ceramic hydroxyapatite also allows human tissue to re-grow around the new joint just as it did around the original bone.

With advances in teaming these brittle materials with more ductile polymers and metals, the range of their use will continue to expand.

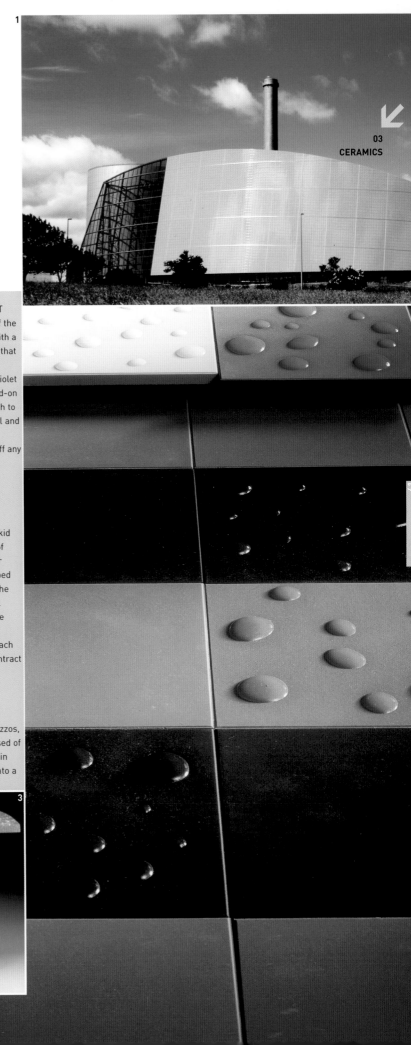

**2**

### 1 AGROB BUCHTAL
KERAION HYDROTECT
The exterior panels of the building are coated with a self-cleaning surface that works through photocatalysis. Ultraviolet light causes the baked-on titania within the finish to react with surface soil and bacteria, allowing rainwater to slough off any dirt.

### 2 ROYAL TICHELAAR MAKKUM
DROPTILES
To make these non-skid floor tiles, each one of which is unique, clear molten glass is dropped onto a ceramic tile. The coefficient of thermal expansion is the same between the two materials, allowing each tile to expand and contract without cracking.

### 3 DOUG FITCH
LEONICA TABLE
Several colored terrazzos, in every case composed of marble chips in a resin binder, are molded into a playful, unique table.

**3**

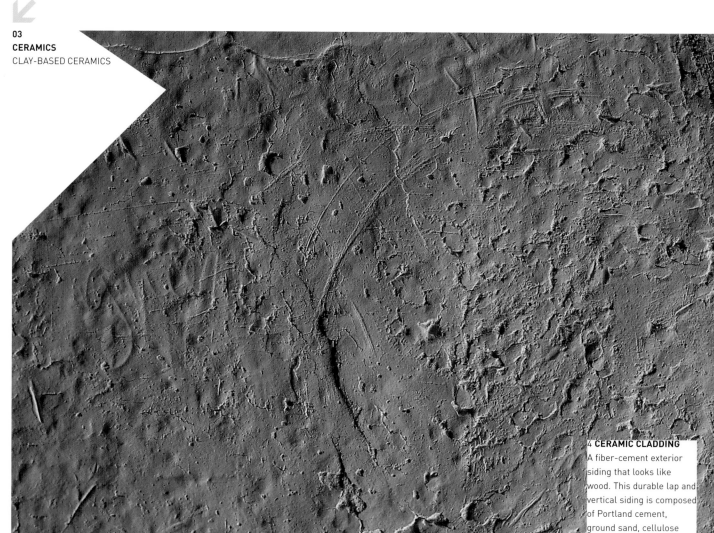

1

**4 CERAMIC CLADDING**
A fiber-cement exterior siding that looks like wood. This durable lap and vertical siding is composed of Portland cement, ground sand, cellulose fiber, additives and water. It is formaldehyde-, fiberglass- and asbestos-free, and non-combustible. 1032-01

**3 REFRACTORY CERAMICS**
Moldable refractory insulation for high thermal-shock applications. 90 per cent alumina/10 per cent silica materials composed of long, high-strength ceramic fibers in an inorganic binder/rigidizer. The ceramics are 100 per cent inorganic and non-combustible. 3695-02

**1 CURVABLE WALLBOARD**
Polymer-modified cement board reinforced with an alkali-resistant fiber mesh. Can be used in both interior and exterior applications around ceilings, beams, columns, arches, archways and walls, and anywhere an evenly curved surface is required. 1177-01

**2 UNFIRED TILE**
Durable porcelain tile with enhanced cleanability containing 95 per cent recycled, unfired raw materials, for areas that require a less expensive tile. Available in a taupe-gray color in two surface finishes: unpolished and with a raised pattern. 48-01

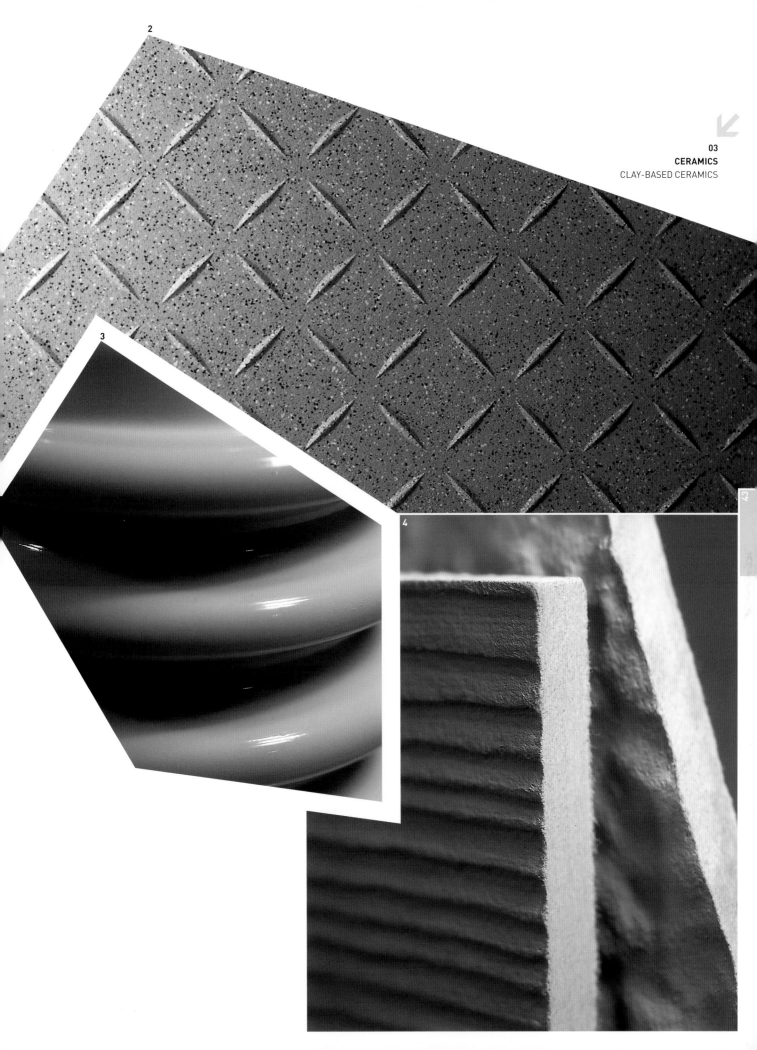

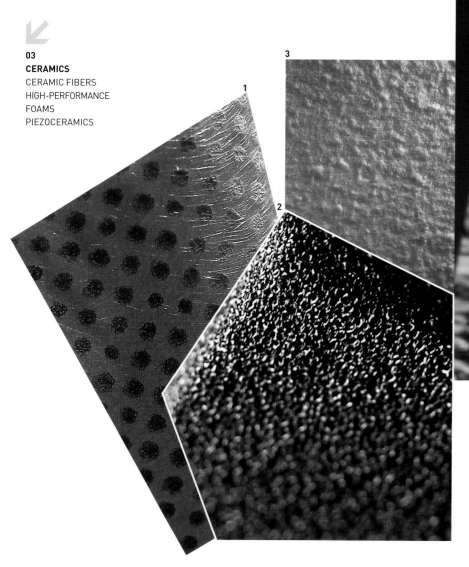

### 6 QUARTZ TEXTILE
Lightweight, flexible and
not prone to shrink or melt
when exposed to fire, this
nonwoven ceramic flame
barrier and thermal-
containment paper is
composed of crystalline
aluminum oxide, silica and
boron-oxide fibers.
3454-02

### 8 SILICA ROPE
Fibers for silica rope used
for high-temperature
packings, seals and cores.
Its coating decomposes at
high temperatures, thus
improving its handling
properties. 2968-07

### 9 PIEZOCERAMIC
Piezoelectric ceramics
fabricated into various
types of devices. They
are made of crystalline
materials (in this case,
of lead zirconate titanate)
that can transform
electrical energy into
sound, motion, force or
vibration when subjected
to a small stress. 4329-01

### 1 ALUMINA NONWOVEN TEXTILE
Nonwoven ceramic flame
barrier and thermal-
containment paper. The
paper is comprised of
crystalline aluminum
oxide, silica and boron-
oxide fibers, is lightweight
and flexible, and does not
shrink or melt when
exposed to fire. The dots
maintain the paper's
integrity. 3454-02

### 2 CERAMIC FOAM
This ceramic (silicon
carbide) foam's open-
celled, three-dimensional
latticework structure
made of interconnected
ligaments makes it strong,
lightweight, porous, and
fracture- and thermal
shock-resistant. 129-01

### 3 ALUMINA-SILICA NONWOVEN TEXTILE
Long, spun, high-purity
alumina-silica fibers are
formed into a highly
flexible sheet to create this
lightweight, resilient,
refractory material, which
contains no asbestos, has
good dielectric strength
and low thermal
conductivity, and is
chemical- and corrosion-
resistant. 174-02

### 4 QUARTZ TEXTILE
Textiles made from
high-purity (99.9 per
cent silicon dioxide)
quartz fibers. At high
temperatures, pure quartz
crystals have significant
tensile strength, and the
diameter of the fibers that
can be attenuated from
them is specifiable.
2975-01

### 5 CERAMIC FIBER
A strong, continuous-
filament ceramic fiber
made of metal oxides.
With low thermal
conductivity, porosity,
elongation and shrinkage,
this fiber is also resistant
to thermal shock. 3454-01

### 7 ZIRCONIA FELT
A mechanically interlocked
felt composed of 100 per
cent ceramic fibers (yttria-
stabilized zirconia fibers
0.0002 in./4–6 µ in
diameter) with no binders.
It is designed for use as
thermal insulation in
corrosive environments
at temperatures above
3,500°F/1,930°C. 3695-01

5

6

7

8

9

**1 UNGLAZED TILE**
Unglazed quarry tile for indoor and outdoor use. Its tight die-skin surface is moisture-, soiling-, impact- and abrasion-resistant.  24-01

**2 INSULATING BOARD**
Insulating millboard that is smooth and flat, strong, thin and dense but lightweight. It is made from ceramic fiber, clay and inert fillers, with a small amount of organic binders added to increase handling strength.  174-03

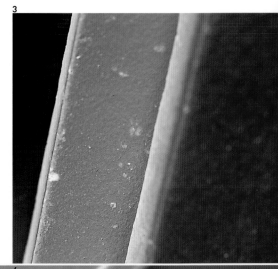

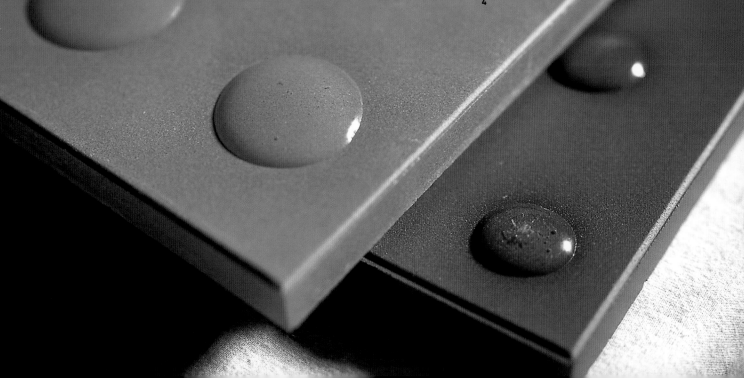

3 **DECORATIVE WALL TILE**
Decorative earthenware wall tiling enhanced with unique sculptural surface effects. Various designs are available, including 'swell', 'pierce', 'pull' and 'press', as well as plain tiles. 4774-03

4 **DECORATIVE FLOOR TILE**
Matte-glazed ceramic floor tiles, the surface of which is decorated with an irregular pattern of adhered glass 'droplets'. 4774-01

5+6 **PORCELAIN TAPE**
Tape-casting a slurry of porcelain ceramic powder produces these flexible, translucent porcelain foils that are thin (0.0008–0.08 in./200 µ–2 mm) in the 'green', pre-sintered state. 4134-01

7 **DIGITAL-IMAGE TILE**
Glazed tiles with custom-designed, digitally reproduced images in full color in monotone, duotone, tritone or quadtone, which are waterproof and slip-, abrasion- , ultraviolet-radiation- , frost-, and thermal shock-resistant. 1406-01

↙
04
**GLASS**

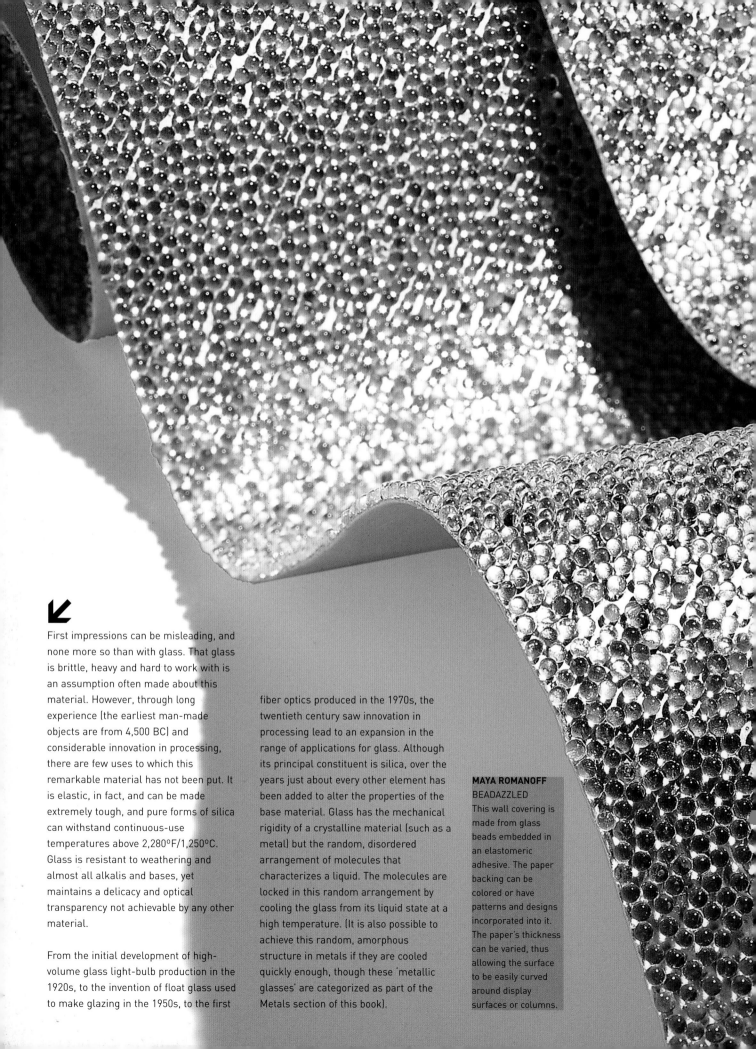

First impressions can be misleading, and none more so than with glass. That glass is brittle, heavy and hard to work with is an assumption often made about this material. However, through long experience (the earliest man-made objects are from 4,500 BC) and considerable innovation in processing, there are few uses to which this remarkable material has not been put. It is elastic, in fact, and can be made extremely tough, and pure forms of silica can withstand continuous-use temperatures above 2,280°F/1,250°C. Glass is resistant to weathering and almost all alkalis and bases, yet maintains a delicacy and optical transparency not achievable by any other material.

From the initial development of high-volume glass light-bulb production in the 1920s, to the invention of float glass used to make glazing in the 1950s, to the first fiber optics produced in the 1970s, the twentieth century saw innovation in processing lead to an expansion in the range of applications for glass. Although its principal constituent is silica, over the years just about every other element has been added to alter the properties of the base material. Glass has the mechanical rigidity of a crystalline material (such as a metal) but the random, disordered arrangement of molecules that characterizes a liquid. The molecules are locked in this random arrangement by cooling the glass from its liquid state at a high temperature. (It is also possible to achieve this random, amorphous structure in metals if they are cooled quickly enough, though these 'metallic glasses' are categorized as part of the Metals section of this book).

**MAYA ROMANOFF**
BEADAZZLED
This wall covering is made from glass beads embedded in an elastomeric adhesive. The paper backing can be colored or have patterns and designs incorporated into it. The paper's thickness can be varied, thus allowing the surface to be easily curved around display surfaces or columns.

As in the field of metals, where it seems that all the great leaps forward in properties have already been achieved, improvements in the glazing and woven-glass fields have come in smaller and smaller increments despite a wealth of recent research. The most significant innovations have come from the synergistic effect of teaming glass with other materials, such as coatings for reducing heat transmission, self-cleaning coatings for window glazing, or decorative and functional interlayers sandwiched between sheets of safety glass. However, the recent use of glass-based aerogel foams is evidence of how dramatically alterations in the form of a material can lead to new properties and applications, such as collecting dust samples from the tail of a comet.

Several types of advanced glazing systems are now available to help control heat loss or gain. Advanced glazings include double- and triple-pane windows, with coatings such as low-emissivity (low-e), spectrally selective, heat-absorbing (tinted) or reflective; gas-filled windows; and windows incorporating combinations of these options. Self-cleaning glazing utilizes a very thin, clear, titania ($TiO_2$)-based coating baked onto the outside surface that when activated by low levels of sunlight breaks down organic dirt, which is then sloughed off by the rain; the possible energy savings from this new development are immense.

Polyvinyl butyral (PVB), the polymer interlayer widely used in automotive windshields and in bullet-proof, sound-deadening and solar-control laminated glass, is increasingly being used as a design tool for creating colored and patterned glazing. The ductility and toughness of the interlayer also give the laminated glass a much greater versatility; bomb-proof glazing

1

installations use this form of lamination, securing the sheets solely at overlapping pieces of the PVB layer rather than through the much more brittle glass sheet.

The reawakening of interest in aerogels has shown how advances in processing can renew interest in an old material. These incredibly low-density glass-foam solids were discovered in the 1930s and developed as additives in cosmetics and paints, but recent advances have led to the production of large solid volumes of the foam and applications that utilize its tremendous insulative and filtrative properties. Aerogels can be made in which 99 per cent of their volume is air. This makes them superior thermal insulators to the extent that only very thin sections are required for blocking extreme temperatures. Able to support more than four thousand times their own weight (though conversely rather brittle and weak), their application as a translucent, structural, highly insulating architectural material cannot be long in coming.

1 **FTL**
TENSILE
FIBERGLASS
TEXTILES
Coated-fiberglass
tensile architecture
used to create
performance spaces
at the World Financial
Center Winter
Performance
Structures,
New York City, 1987.

2 **ANDY CAO**
RED BOX
A grass wall and
recycled medicine
bottles fused
together to form tiles
were used to created
this installation at the
American Academy
in Rome.

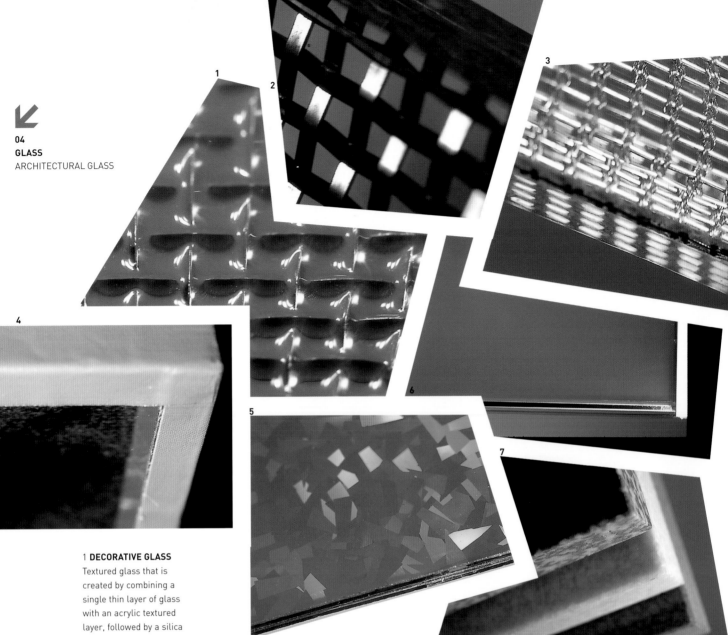

1

2

3

4

5

6

7

**1 DECORATIVE GLASS**
Textured glass that is created by combining a single thin layer of glass with an acrylic textured layer, followed by a silica coating for abrasion resistance. Proprietary technology creates geometric, textural or combination surfaces.
49-02

**2 SAFETY GLASS**
Flat metal materials, fabrics and woven-wire cloth are laminated between panels of various types of tempered and patterned glass; for architectural and design applications, lighting, and interior and exterior glazing systems.
1965-05

**3 SAFETY GLASS**
Decorative glass for interior and exterior architectural glazing. Composed of a sheet of woven copper mesh sandwiched between two layers of clear acrylate film, which is bonded to two layers of tempered glass. Can be made with any colored glass.
1965-08

**4 HIGH-STRENGTH GLAZING SYSTEM**
A glazing system composed of self-supporting glass channels within a custom-designed aluminum frame. The cast-glass channels have high structural strength and are contained within the aluminum perimeter frame.  4283-01

**5 HOLOGRAPHIC GLASS**
Laminated glass that has decorative holographic designs. This window-glazing material combines the safety, security and sound control of laminated glass with transmissive and reflective designs and effects created using holographic techniques.
4378-01

**6 ELECTROACTIVE GLASS**
Electrically operated architectural glass for controlling vision. It is constructed of two clear or tinted sheets of glass laminated with a liquid-crystal film between two plastic interlayers that turns from clear to opaque when a current is applied.
4477-01

**OMNIDECOR**
DECORFLOU
Glass, back-painted and
backlit, is used here for
retail display. Traditionally
employed for interior and
exterior architectural
cladding, abrasion-
resistant and anti-
reflective, it is ideal for
showrooms.

**7 GRAFFITI-PROOF
CLADDING**
Glass architectural
cladding material with
a high-gloss surface,
for interior and exterior
applications. Though much
lighter than stone, it can
be formed into thin but
strong flat and curved
panels that are as strong
as much thicker pieces
of stone.  1413-01

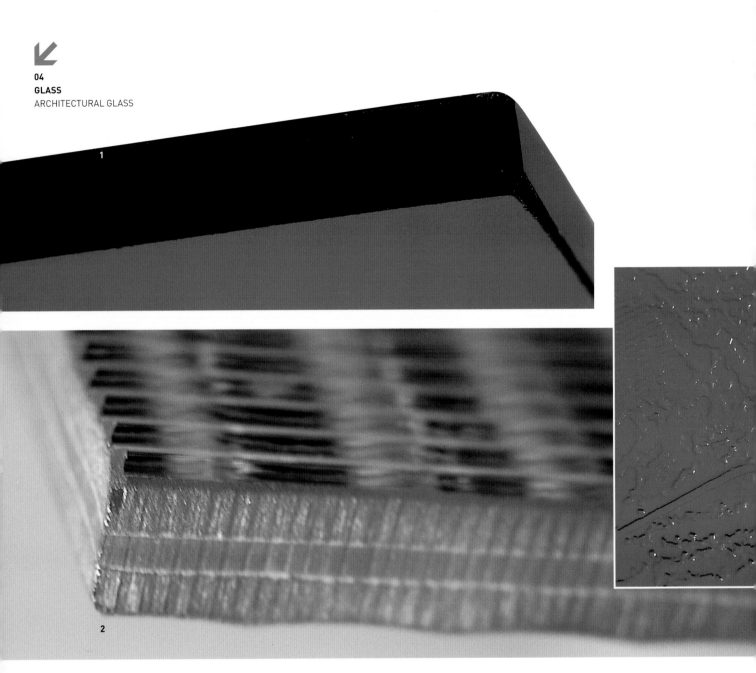

1 **UNIDIRECTIONAL-VIEW GLASS**
By looking like a mirror in a well-illuminated room but working like an ordinary tinted window from the other side, this window glass permits unobtrusive observation in interior applications. 1459-04

2 **LAMINATED GLASS**
Sheet glass of varying colors and textures laminated together. 36-03

3 **ETCHED GLASS**
Sandblasted glass with delicately scaled directional patterns in ten standard etched-glass patterns. Both clear and opaque finishes are available.  18-01

4 **DECORATIVE GLASS**
A hand-crafted architectural glass with the look and texture of more costly kiln-formed glass. It can be made of plate, tempered or laminated glass in clear, frosted and color finishes. 18-02

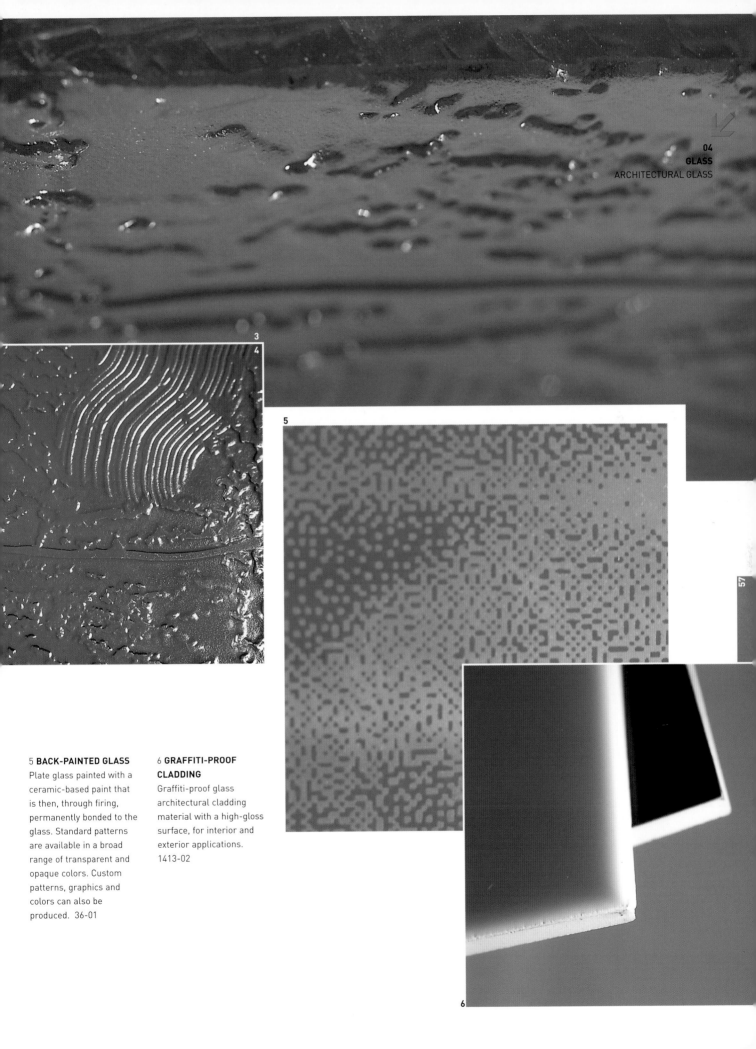

3

4

5

57

**5 BACK-PAINTED GLASS**
Plate glass painted with a
ceramic-based paint that
is then, through firing,
permanently bonded to the
glass. Standard patterns
are available in a broad
range of transparent and
opaque colors. Custom
patterns, graphics and
colors can also be
produced.  36-01

**6 GRAFFITI-PROOF
CLADDING**
Graffiti-proof glass
architectural cladding
material with a high-gloss
surface, for interior and
exterior applications.
1413-02

6

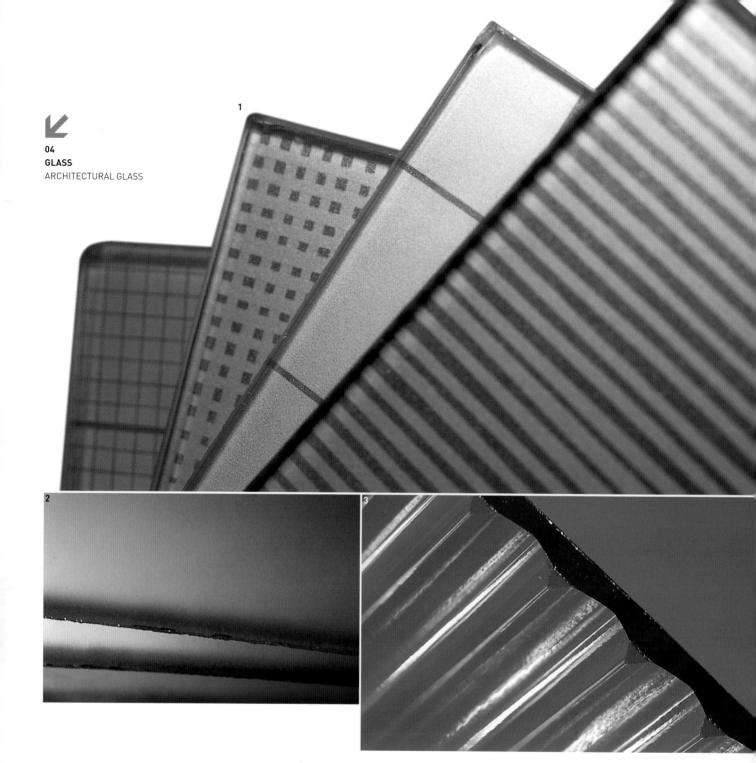

**1 LAMINATED SAFETY GLASS**
A durable, low-maintenance 9/32 in./ 7 mm laminated safety glass consisting of two thin layers of clear 1/8 in./3 mm annealed float glass permanently bonded with a 0.04 in./9 mm thermoplastic interlayer. 49-01

**2 COLORED MIRRORED GLASS**
Luminous mirrored glass that is colored throughout. Silica and metal oxides give the glass its color. It is fabricated by chemically etching colored or clear float glass on one surface and then silvering the non-etched side. 1252-02

**3 PRISMATIC GLASS**
Precision-rolled, patterned glass with flutes impressed at an angle, thus creating a prismatic effect. The image viewed through the glass changes depending on the angle of view and whether the glass is used horizontally or vertically. 1252-03

**OCEANSIDE GLASS TILE**
MOSAIC GLASS TILE
Water flows gently down
this mosaic wall of
1 in./2.54 cm-square
glass tiles in cobalt blue.
The divers are frozen in
motion, perpendicular to
the wall at the top and
slowly arcing down to
become parallel to the
wall at the bottom.

1

2

3

### 1 COLORED GLASS

Decorative coated glass. A lead- and cadmium-free varnish produces the metallic-colored, decorative surface effects in this hand-coated glass. Custom colors are available, as well as nine standard ones. The coating may be applied to flat or curved surfaces.  1505-03

### 2 COLORED TEXTURED GLASS

Translucent, textured, colored glass featuring a pressed wrinkle design on one side and ribs on the other.  1648-01

### 3 GLASS TERRAZZO

Terrazzo tiles and slabs composed of crushed mosaic glass and granite mixed in a polyester-resin matrix and backed with crushed white marble. 1679-06

### 4 CONTROLLED-VISION GLASS

Laminated glass composed of two panels of annealed float glass sandwiching a plastic film. Depending on the angle from which it is viewed, the glass changes its appearance from transparent to translucent or the reverse.  1965-01

### 5 DECORATIVE LAMINATED GLASS

Flat metals, fabrics and woven-wire cloth are laminated between panels of various types of tempered and patterned glass. The result has applications in the fields of lighting, design and architecture, for instance in interior and exterior glazing systems.  1965-05

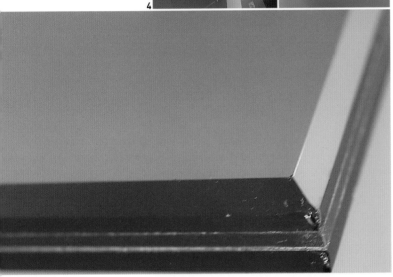

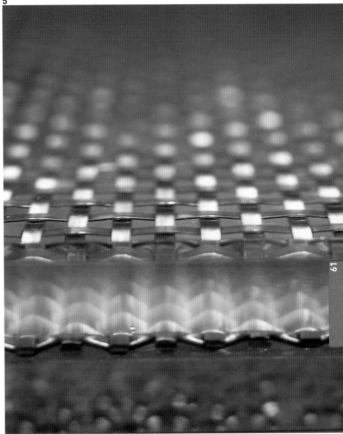

**6 CUSTOMIZABLE DECORATIVE GLASS**

Heavily patterned, textured glass for architectural glazing systems that reflects light and images from multiple angles. It can be custom-silvered and is available in four hundred custom patterns and a wide range of colors. 1965-02

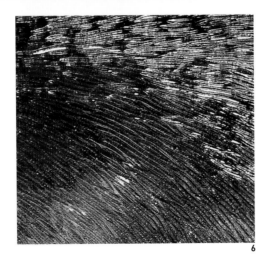

6

**7 LAMINATED INTERIOR GLASS**

A film of decorative rice paper is sandwiched between two layers of glass to create an even, diffuse light. Available in four types, this glass is suitable for interior applications. 1965-06

7

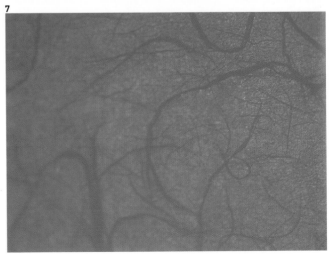

**2 DICHROIC GLASS**
Glass that has been vacuum-coated with stacks of very thin layers of metal oxides creates an interference filter that causes the glass to reflect one color and transmit the complementary one, depending on the number of layers in the stack. 2663-01

**3 RECYCLED GLASS**
Custom-textured glass for architectural applications. 100 per cent recycled glass is annealed before being cast into custom-designed ceramic molds and then tempered to increase strength. 2721-01

**4 RECYCLED GLASS TILE**
Decorative tiles made of 100 per cent recycled glass. They are available in a variety of sizes, shapes, designs, patterns and colors, and can be used for walls, floors, borders and inserts. They come in sheets that measure 0.96 by 0.96 ft/ 0.3 by 0.3 m each. 2686-01

**6+7 RAW GLASS**
Designed for kilnworking, this line of products is based on the following criteria: compatibility (the likelihood that the glass will fully fuse in most kilns above 2,642ºF/1,450ºC), color stability and resistance to devitrification. 3591-01

**1 DECORATIVE LAMINATED GLASS**
After this glass has been vacuum-coated to make it refractive, it is laminated with patterned, tempered float glass and/or textured glasses using clear or colored interlayers. It is used for architectural and design applications, lighting and interior and exterior glazing. 1965-07

**5 FABRIC-BACKED GLASS**
After being backed with paint, paper or fabric, this glass is epoxy-coated. It can be used for wall coverings, table- and countertops, tiles and decorative pieces. 2722-01

1

2

3

4

9 **RECYCLED GLASS TILE**
These tiles, which are made from 100 per cent recycled post-consumer/post-industrial glass, can have a textured or regular finish, and have a higher breaking strength and abrasion-resistance than porcelain-type paver tiles. They are available in eleven colors for applications such as countertops.  3081-01

8 **DECORATIVE GLASS TILE**
The holograph-like surface effects on these decorative glass tiles are created by decorating clear sheets of glass with several layers of a glass-based paint. 4127-01

1311 - 02F

**04**
**GLASS**
ARCHITECTURAL GLASS

**1 LAMINATED-GLASS SYSTEM**
Laminated-glass interlayer system consisting of patterns and metallic colors to create an array of effects. The interlayers are made with heat- and light-stable pigments and are shatterproof, sound-damping and UV-screening. 4684-02

**2 METAL-COATED GLASS**
Thermal arc-spraying various color-producing metals – such as aluminum, bronze, iron, steel and copper – onto either flat or carved glass gives this product its customizable colored surface. 4293-01

**3 SATIN-FINISHED GLASS**
The uniform, soil-resistant surface of this satin-finished glass transmits a large amount of soft, diffused light. 4491-01

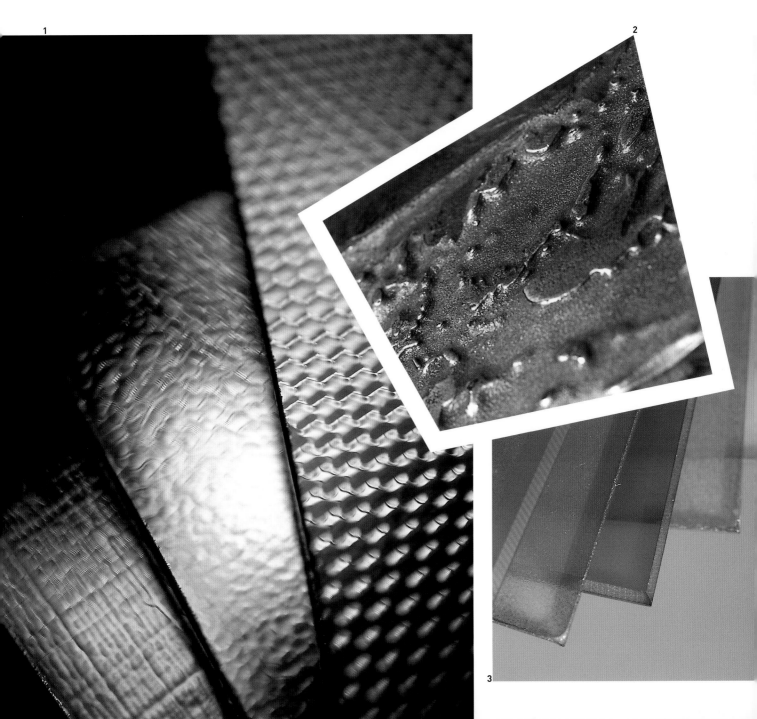

1

2

3

**4 LAMINATED DUAL-COLOR GLASS**

Glazing for architectural applications. Double-sided glazing that appears white or colored depending on the angle of light. Nonwoven gauze matting sandwiched between one pane of colored and one of clear glass produces a white translucent surface. 4864-01

**5 GLASS PANELING**

Aluminum honeycomb sandwiched between 0.16 in./4 mm panes of toughened glass joined together using clear epoxy resin, thus creating a 1.06 in./27 mm-thick panel. 4564-02

**6 GRAFFITI-RESISTANT TILE**

These textured glass-ceramic tiles are resistant to acid, lye and scratches (matte surface only) and allow for easy graffiti removal, making them useful in both interior and exterior settings. The tiles are available in forty color variations and either high-polish or matte finishes. 4659-01

**7 LAMINATED-GLASS SYSTEM**

Interlayer system consisting of nine patterns and fifty-four colors to create an array of effects. Colors and patterns can be used alone or in combinations of up to four layers to create more than eight hundred translucent colors. 4684-01

4

5

6

7

**1 WOVEN-GLASS FIBER**
Fiber used as an interior wall covering that is treated with a natural starch binder for dimensional stability. The textile exceeds Class A fire and toxicity ratings, will not shrink or stretch, and is breathable. 3779-01

**2 COATED-GLASS FIBER**
Coating applied to woven glass-fiber fabrics for tensile membranes. Physical vapor deposition coats both sides of the e-glass fiber with a thin layer of aluminum. 4998-01

**3 GLASS-FIBER COMPOSITE GRATING**
Molded industrial grating composed of fiberglass-reinforced plastic. The fibers are fully embedded in any one of several resins, typically polyester or vinylester. Custom resins are also available, as is a fire-retardant option. 188-03

**4 WOVEN GLASS**
A wall covering woven from 100 per cent glass yarn that is stable, abrasion- and stain-resistant, nonflammable, decay-proof and lightfast. 79-01

**5 WOVEN THREE-DIMENSIONAL SPACER TEXTILE**
A sandwich panel that is both high-strength and flame-retardant. This glass-fiber fabric is woven to produce two parallel sides with reinforcing connecting fibers and is impregnated with a phenolic resin. 91-10

**6 WOVEN FIBERGLASS FABRIC**
Fabrics for a wide variety of window treatments, including decorative vertical blinds, sun-screening roller-shades and sun protection for building exteriors. 1786-01

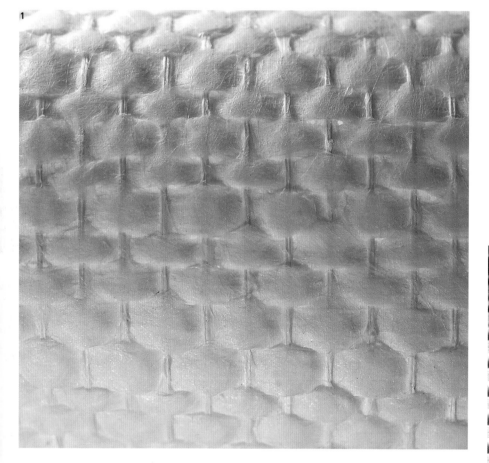

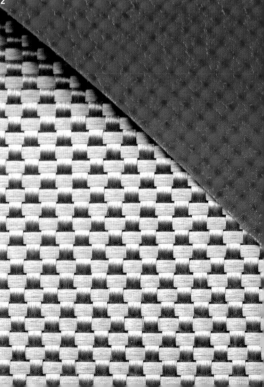

3

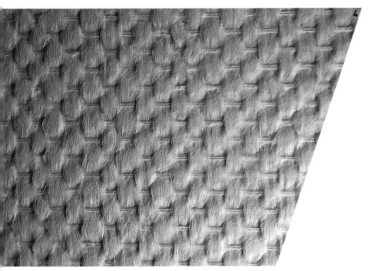

4

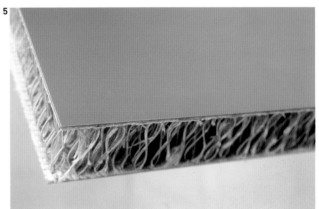

5

6

67

**04
GLASS**
GLASS FIBER

**2 GLASS-FIBER SLEEVING**
Thermal-protection silica sleeving for metal tubing, hose lines and cables. The sleeving's coating decomposes at high temperatures to improve handling properties during installation. 2968-03

**3 NONWOVEN COMPOSITE REINFORCEMENT**
A core material composed of a 'random-laid', non-continuous polyester nonwoven, combined with 50-per-cent-by-volume expanded plastic microspheres and a glass mat. 2492-02

**4 GLASS-FIBER INSULATION**
This pre-shrunk, dimensionally stable, high-temperature, silica-based insulation was designed as a replacement for asbestos; applications include furnace covers, stress-relief blankets and resin impregnation. 2968-01

**5 EXTREME TEMPERATURE-RESISTANT GLASS FABRIC**
A specially treated, premium-grade fiberglass fabric that can withstand temperatures up to 1,200°F/649°C: well above the limit of standard fiberglass cloth. 2968-06

**6 LIGHTWEIGHT ARCHITECTURAL FABRIC**
Coating a glass fabric on both sides with a base coat results in this lightweight, high-strength, flexible, weldable, translucent, fire-resistant architectural fabric. 3088-02

**1 WOVEN COMPOSITE REINFORCEMENT**
Fiberglass reinforcement for FRP composites. Structural e-glass is made from strands of continuous glass filaments that are plied and twisted into yarns. This material offers a unique combination of physical and electrical properties. 1786-02

**JOHNS MANVILLE**
WALL COVERING
A wall covering woven
of fiberglass, a material
that lends itself to a wide
variety of finished textural
patterns. After being
woven, the textile can be
flattened or embossed.

**04**
**GLASS**
GLASS FIBER

**1 NONWOVEN ACOUSTIC
TEXTILE**
A thin, fire-retardant,
textile made of cellulose
and glass fibers. It can
be glued to the back of
perforated ceiling tiles
(of metal, wood, plastic
or gypsum) by means of
an adhesive that is already
applied to the fabric.
3136-01

**2 INTUMESCENT COATING**
A coating that insulates
electric cables during a
fire. Spraying or brushing
on to all visible surfaces
of a cable bundle produces
an insulating foam layer
that, when exposed to
heat, expands to sixty
times its original
thickness.  4089-01

**3 THREE-DIMENSIONAL
SPACER FIBERGLASS
TEXTILE**
A fabric consisting of two
bi-directional, woven-
fiberglass (e-glass)
surfaces that are
mechanically connected
with vertically woven
piles in a way that leaves
a predetermined space
between the two surface
layers.  3212-01

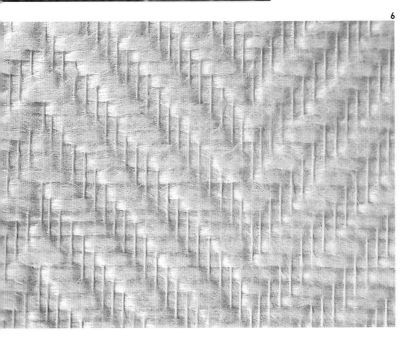

4

5

6

71

4 **WEATHER-RESISTANT COATED-GLASS FIBER**

A silicone-coated fiberglass for use in fabric architecture. This translucent fabric, composed of 56 per cent fiberglass and 44 per cent silicone, is flame- and UV-resistant, has high tensile and tear strength, is flexible and retains elasticity. 4247-01

5 **COMPOSITE REINFORCEMENT**

A new material for manufacturing lightweight, stiff FRP laminates. It consists of a glass-fiber base with thermoplastic microballoons for volume. This base is mechanically compressed and held under compression by a stitch-bonded web of polyester fibers. 4143-01

6 **DIMENSIONALLY STABLE WOVEN GLASS**

Glass fiber treated with a natural starch binder for dimensional stability and used primarily as an interior wall covering. The textile exceeds Class A fire and toxicity ratings. It will not shrink or stretch and is breathable. 3779-01

### 7 GLASS CONCRETE

Recycled glass embedded in a proprietary concrete mixture without the use of a resin binder. Developed by Columbia University in New York, this is the first cement suitable for use with glass. 4853-01

### 8 GLASS FOAM

Cellular glass foam for pipe-lagging insulation. Composed of 100 per cent silica, without any binder, in densities from 8 to 10.2 lb/ft$^3$/128 to 150 kg/m$^3$). It can withstand temperatures ranging from -450 to +900°F/ -268 to +482°C. 4597-01

### 1 MOSAIC TILE

Custom-designed mosaic glass tiles. To ease installation, stained glass and mirror pieces are positioned on either a mesh backing or a paper front. Border tiles are 1 linear ft/10 by 30 cm, and field tiles are 1 sq ft/ 30 by 30 cm. 4757-01

### 2 GLASS MICROSPHERES

These clear, solid and hard glass beads have become an integral part of our highways by making roadside signs more retroreflective. 4198-01

### 3 EFFICIENTLY LIGHT-TRANSMITTING INSULATION GLASS

Light-diffusing and insulating panels that provide shadowless light with true color rendition. They are composed of hollow capillary tubes that lie between tissue layers of glass fiber. This stratum is hermetically sealed within float glass. 3639-02

### 4 RECYCLED GLASS GRANULATE

These lightweight, expanded glass granules, which have a uniform, fine, porous structure and a compact, closed surface, are durable and thermally insulating, have high compressive strength and can be recycled. 4119-02

### 5 GLASS FOAM

A building material made of cellular glass that is durable, lightweight and rigid. This glass foam is composed of millions of completely sealed glass cells, each forming an insulated space. It is resistant to water and water vapor and is non-corrosive, non-combustible and highly insulative. 4494-01

### 6 AEROGEL

Extremely lightweight, porous solids which are highly insulating. Because of their exceptionally high surface area, they also effectively remove or recover pollutants from air and water. 4131-01

↙

Few metals currently used in design, construction and architecture could be classed as new. Often the best solution to a materials problem will be a metal first discovered and mined hundreds, sometimes thousands, of years ago. Of course improvements in the ultimate performance of the metal will have been achieved by 'alloying' with small amounts of other metals or even non-metals such as carbon, nitrogen or silicon, giving better corrosion resistance, greater toughness, magnetic – or non-magnetic – properties, improved scratch resistance or easier weldability.

In fact, there are very few applications for metals in which the pure element is used, and, more often than not, there are likely to be three or more added elements. This type of metallurgy has been taken to the ultimate level with applications such as jet-engine turbine blades that often incorporate thirteen or fourteen additional elements into the nickel metal to maintain performance in extreme environments.

Heat treatment of a metal can also have a significant effect on its properties. For instance, many aerospace aluminum alloys are designated by their heat-treatment regime. Thus, although the original metal may be as old as (and even originally part of) the hills, developments in alloying and heat-treating have given us continuous improvements in all aspects of their properties.

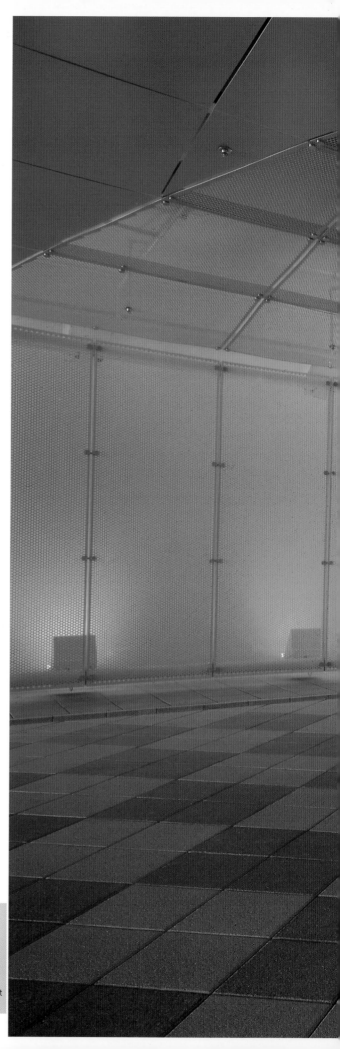

**MORADELLI**
PANEL
Perforated steel sheets are used for wall and ceiling panels, allowing the transmission of light and flow of air.

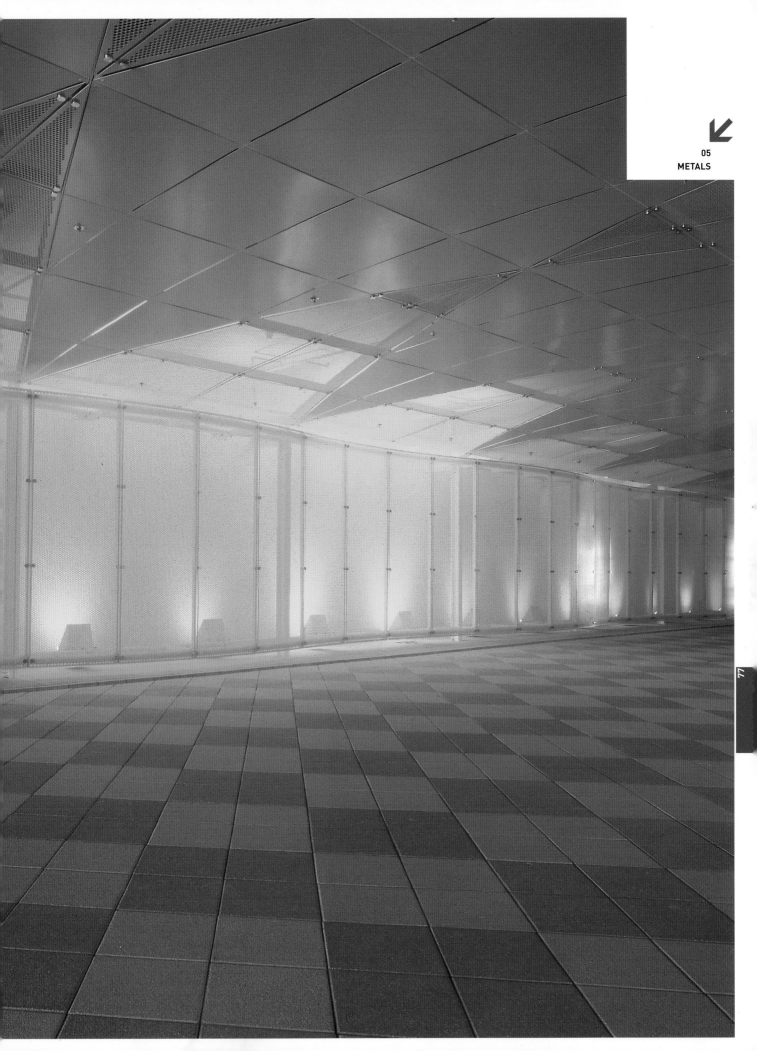

Titanium can be considered an interesting alternative to this rule. It is a young metal, only used in commercial quantities since the 1950s, mainly due to the difficulty of refining and processing it; for similar reasons, it is also very expensive. A truly high-performance material, it is used widely in aerospace applications where high strength-to-weight ratios and stiffness properties are paramount. It also has exceptional corrosion resistance. Its use as cladding in architecture has brought it to the forefront of designers' minds, though perhaps this is not the most apposite demonstration of its properties; there is a reason why titanium alloys are used for the load-bearing struts in an aircraft and yet aluminum is preferred for the skin.

In attempts to increase the strength-to-weight ratio of metals, numerous forms have been developed, of which corrugation, honeycomb structures, perforation (page 103/3), cambering (page 104/2) and foaming (page 96/8–10) are four. Of these, metallic foams, in particular aluminum foams, have created interest among designers and engineers alike due to their extremely low density, as well as their unique aesthetics. Created simply by passing a gas into the heated liquid aluminum (analogous to blowing into a milkshake with a straw), the resulting foam is extruded or cast, with the cell or bubble size, surface integrity and density easily controllable. It must be noted, however, that despite

the originally developed application for this remarkable material – as a firewall and energy-absorbing panel put between the engine and occupants in cars – there have, so far, been no successful attempts to fully exploit its distinctive properties.

From automotive interiors to electrical appliances, it is clear that metal continues to retain an aura of strength, durability and performance. However, the attempt to recreate this aura visually with metallic coatings, while utilizing a lower-cost substrate material like a polymer, has so far been unsuccessful without the use of metal in some form. Thus coatings, amalgams, films and pigments have seen significant development in attempts to make non-metals (normally plastics) look metallic. As well as the many electroplating, vapor-deposition and thermal-spray processes for laying down a layer of metal, of particular interest of late have been coatings and castings that combine fine metal particles with a polymer carrier. In the case of coatings (page 83/6), a cold-spray process is used to deposit a layer that mimics a burnished or polished metallic surface without the need for heat, a plating bath or expensive vacuum-deposition equipment. This type of polymer-and-metal-powder composite may also be cast (page 103/4), thus producing intricate, textured large-scale surfaces that can be re-polished if scratched and are also relatively lightweight.

There has been significant interest also in responsive materials, with metals in particular demonstrating a range of useful responses to external stimuli. The shape-memory metals (page 104/6), predominantly alloys of nickel and titanium, exhibit memory of their original shape following deformation when exposed to an increased temperature. These have found application in auto-off switches in electric kettles, shrink-to-fit seals and – perhaps best known – unbreakable spectacle frames. Other examples of responsive materials are the magneto-rheological or ferro fluids (page 100/10), fine metal filings in a paste that instantly change from malleable liquid to rigid solid on the application of a small magnetic field. It is also possible to turn any solid surface into a speaker simply by attaching a magnetorestrictive metal; electrical impulses to this metal will cause it to change shape and then return at high frequency, vibrating the surface it is in contact with and causing that surface to emit sound.

Overall, despite the limited increments in performance currently experienced by most metals and their alloys, new forms, new applications and newly discovered properties will continue to keep this category in the forefront of innovative solutions for design and engineering alike.

**MABEG (DESIGN AFAIRS)**
ALUMINUM STOOL
A single flat sheet of aluminum is first corrugated to increase stiffness, then bent to create the sides of this stool (thereby increasing rigidity and reducing the chances that it will torque), and finally deformed using a high-pressure process to create the seat.

**COBRA**
KING COBRA
The exceptionally high strength-to-weight ratio of titanium, used in the head and shaft of this golf club, reduces its weight, allowing the golfer to achieve greater force when hitting the ball while using less energy.

**05**
**METALS**
COATINGS

**2 INTERFERENCE-COLOR CHROME COATING**
Electrochemically depositing chrome coatings of varying thickness produces the champagne, bronze, blue, gold and black colors of this stainless steel. The coatings do not crack, craze or fade in the sun. 35-01

**3 ELECTRODEPOSITED HIGH-GLOSS TITANIUM COATING**
A uniform coating (0.002 in./0.005 cm thick) of titanium on highly mirror-polished stainless-steel sheet or railing that creates the appearance of brass without needing to be lacquered. 35-02

**4 DECORATIVE TEXTURED METAL SHEET**
Coated embossed sheets with a variety of patterned surfaces that are available in many metals, including stainless steel, aluminum, bronze, brass, copper, cold-rolled carbon steel and titanium. 54-01

**1 COATED LIGHTWEIGHT INTERIOR PANELING**
Decorative aluminum paneling for vertical interior surfaces manufactured in sixteen colors plus custom colors, in ten patterns, and perforated. Stainless-steel versions are also available as a custom option. 3648-01

1

2

4

5

5 **METAL/POLYMER-COMPOSITE INTERIOR CLADDING**
Powdered metal bonded together in a tough fiberglass-reinforced resin matrix. The material is then molded to a range of textures and surface effects that mimic cast-metal surfaces but are lighter than solid metal. 247-01

6 **ARCHITECTURAL CLADDING**
These die-cut architectural metal sheets are available in stainless steel and muntz metal in both satin and polished surface finishes, and in aluminum with colored anodization. The range of textures includes random abrasion, etching, stamping and sandstone. 247-02

7 **DIP-GALVANIZED STEEL**
Steel sheet coated with a corrosion- and oxidation-resistant, 55 per cent aluminum-zinc alloy applied by a continuous hot-dipping process. The sheet can be used both unpainted and painted; applications include roofing, siding and appliance and air-conditioner housings. 767-01

8 **ZINC-BASED EXTERIOR CLADDING**
Durable exterior cladding (zinc alloyed with 1 per cent titanium and 1 per cent copper) for buildings and roofs that is non-flammable and resistant to frost, damp, vapor, UV radiation and rot. 798-01

6

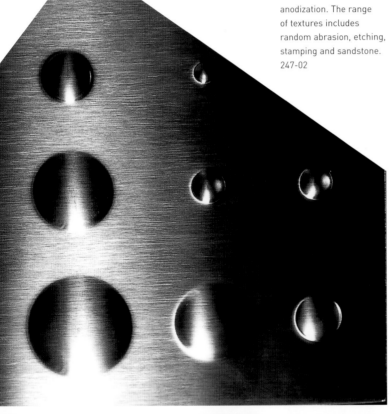

7

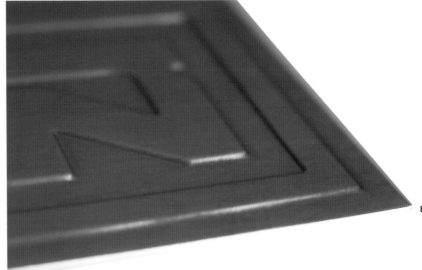

8

81

05
**METALS**
COATINGS

1 **PORCELAIN-COATED STEEL**
Writing surfaces for schools, conference rooms, plants, offices, hospitals and laboratories, made of light-gauge porcelain/ceramic-on-steel. Strong, hard and non-porous, they accept magnetic components and are available in high- or medium-gloss finishes. 1219-01

2 **THIN HIGH-TEMPERATURE INSULATION**
Refractory insulation (e.g. high-performance glass fiber, microporous quartz fibers or metallic mesh) sandwiched between two corrugated, oxidation- and corrosion-resistant, extremely thin metal foils. 1422-01

3 **POLYMER-COATED EXTERIOR SHEET**
Metal sheets that are suitable for outdoor use. Rotational screen-printing applies a polymeric film approximately 0.0004 to 0.0005 in./10 to 12 μ thick to the 100 per cent aluminum sheet. 1475-02

4 **ANODIZED ALUMINUM**
A computer-controlled, two-step anodizing process produces these lightfast, colored aluminum extrusions in blue, green and burgundy that are intended for architectural applications. The process allows for color selection and ensures uniform color from load to load. 1500-01

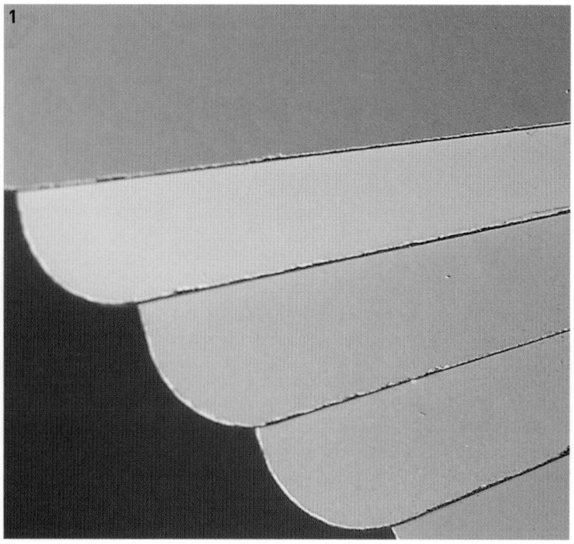
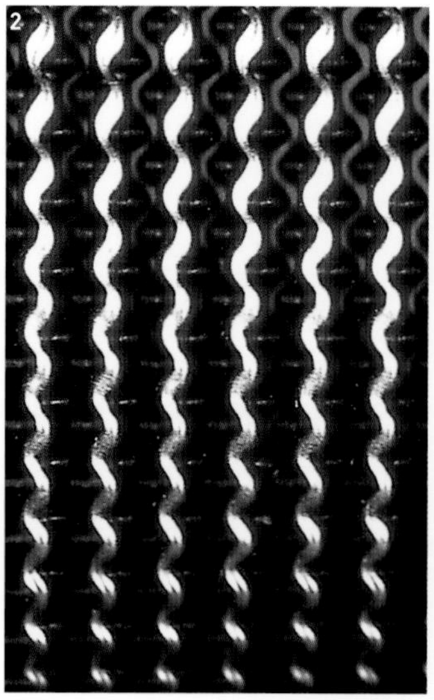
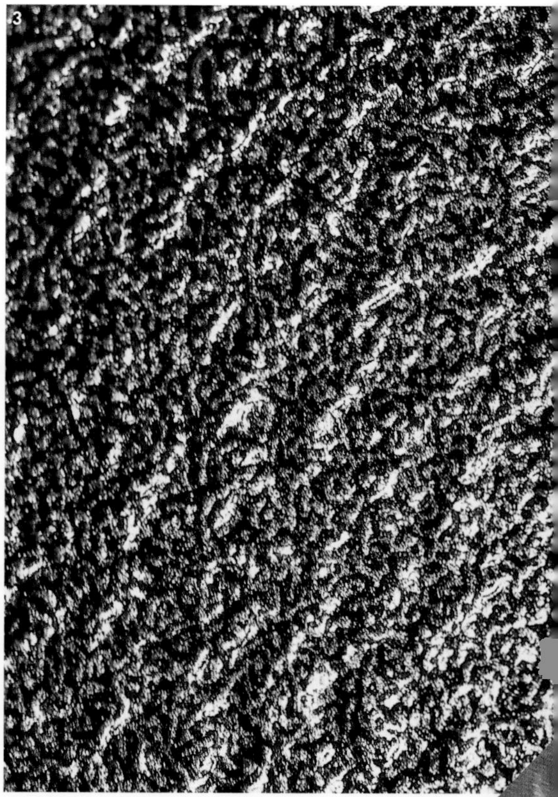

5 **COLD-WORK TEXTURING PROCESS**
A cold-rolling process creates textured patterns on the surface of a wide variety of metals such as stainless steel, aluminum, architectural bronze, copper, brass and titanium. 2693-01

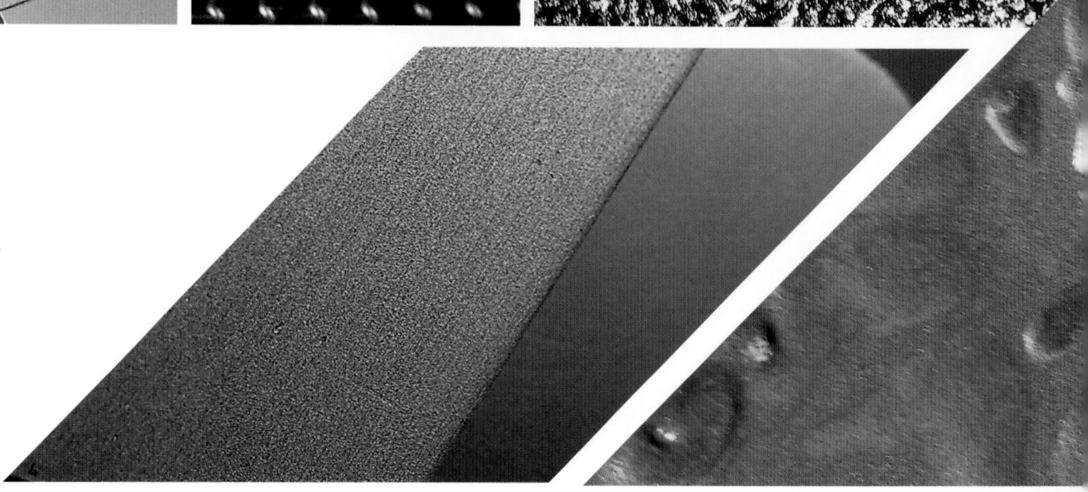
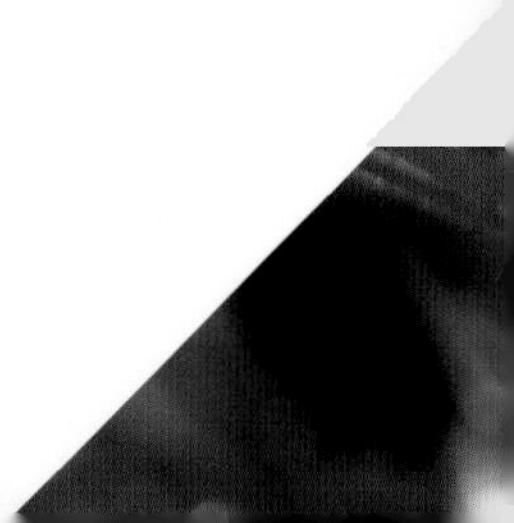

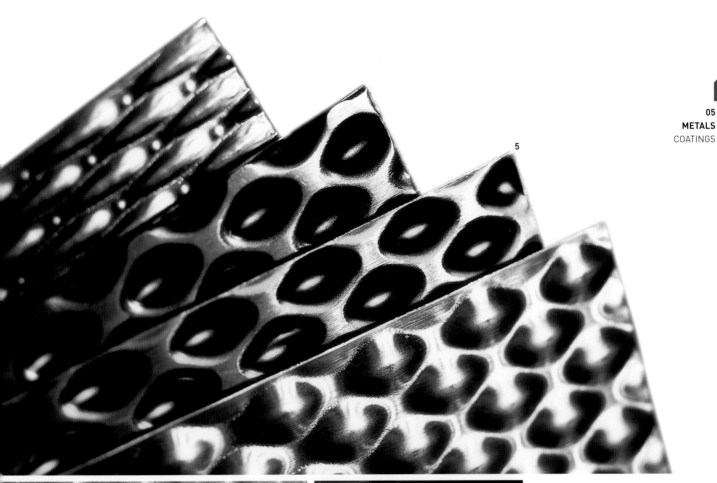

5

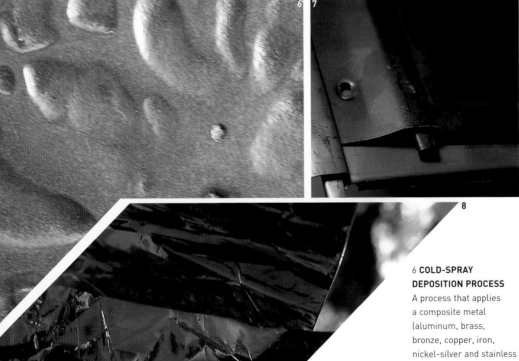

6 7

8

### 7 PRISMATIC STAINLESS STEEL
Stainless steel (T304) permanently colored using a process that draws color from within the steel and enhances its protective chromium-oxide layer. Variations in this layer produce prismatic effects.
4796-01

### 8 METALLIZED INSULATION FOIL
One- and two-sided metallized cryogenic insulation materials with vacuum-deposited aluminum. The materials have an effective separation of layers, low lateral conductivity and low emissivity. Though only 0.0009 in./0.025 mm thick, they provide high mechanical strength.
4575-01

### 6 COLD-SPRAY DEPOSITION PROCESS
A process that applies a composite metal (aluminum, brass, bronze, copper, iron, nickel-silver and stainless steel) to almost any surface (e.g. foam, plastic, gypsum, wood, metal, fiberglass, plaster, ceramic, concrete, terracotta, cardboard).
3076-01

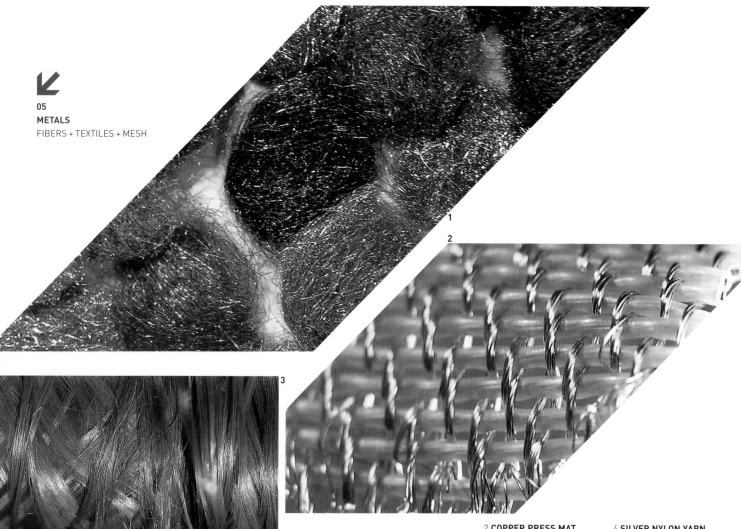

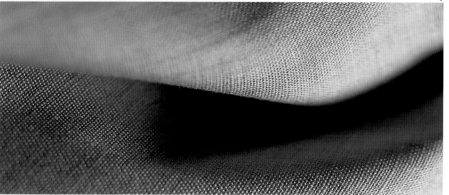

**2 COPPER PRESS MAT**
A highly durable mat that permits efficient heat transfer and quickly regains its original shape after the press cycle.  It is woven in a herringbone twill pattern with a brass warp yarn and a silicone elastomer-encased copper weft yarn that acts as a heat sink.  3907-01

**3 METAL-COATED ARAMID FIBER**
High-strength, lightweight, metal-clad (nickel, copper, tin, silver) aramid fibers that are used for weight reduction in aerospace applications, and braided for electrical and electromagnetic interference (EMI) shielding.  2164-04

**4 SILVER NYLON YARN**
A high-performance silver-coated fiber for textiles. Pure silver is permanently bound to a nylon yarn to create a fiber that is conductive, reflective, anti-microbial, anti-odor, anti-static and thermally insulating. 4625-01

**5 WIRE CLOTH**
Custom metal wire mesh and cloth. Available in aluminum, brass, copper, stainless steel, bronze, iron, silver, tantalum, molybdenum, platinum and gold. Gauges, mesh size and openings are variable, and the sheet size can be custom engineered.  4651-01

**1 DYEABLE DECORATIVE ANODIZED-ALUMINUM FIBER**
Staple fibers consisting of a plain or pigmented aluminum alloy with a special epoxy coating, for shielding against electromagnetic interference (EMI) and electrostatic discharge (ESD). 1466-03

**6 METAL-BEAD DECORATIVE CHAIN SCREENING**
A series of hollow metal beads securely linked by dumbbell-shaped connectors that leave each bead completely flexible yet incapable of kinking, binding or jamming. Available in brass as well as several types of steel and aluminum. 178-02

**7 FILTRATION MESH**
Woven stainless-steel and copper wire mesh of uniform hole dimensions for filtration and screening. Current applications are for architectural drapery as well as powder and fluid filtration. 1575-02

**8 METAL-BEAD DECORATIVE SCREEN**
Designed as decorative screens, these ball-and-bar chains are offered in a wide range of colors, sizes and shapes. The balls are available in nickel-plated steel, gunmetal, brass, aluminum, bronze, stainless steel, yellow brass and zinc. 4927-01

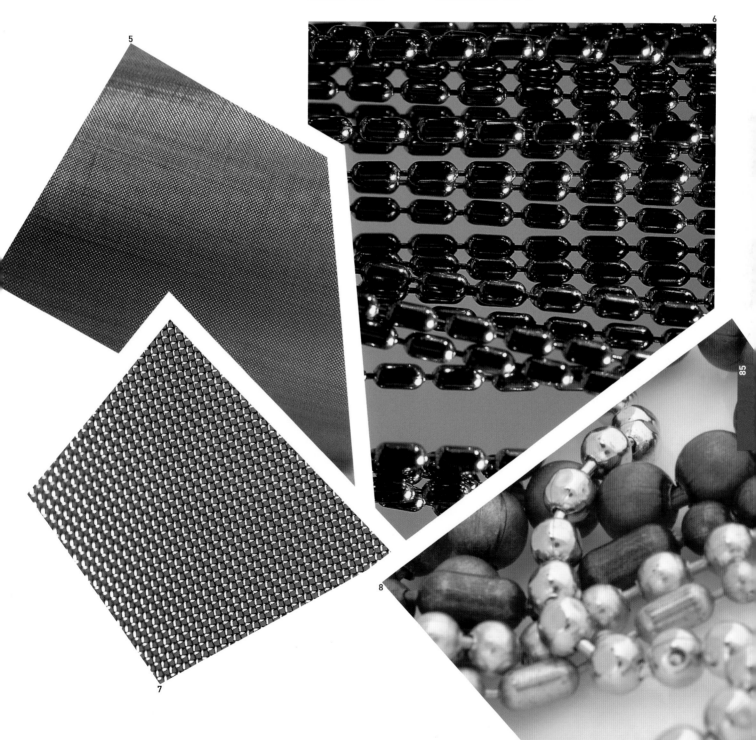

5

6

7

8

### 1 STAINLESS-STEEL MESH

Non-rusting ferrules of tin-coated copper join together lengths of steel mesh made of stainless-steel cables to form a strong, flexible structure. A UV-resistant polyamide coating is optional. 1907-01

### 2 WOVEN-METAL MESH

Fabric manufactured in eight steel or aluminum fabric-weave sizes. Possible metals are nickel, copper, brass and stainless and galvanized steel. Durable acrylic coatings in a wide range of finishes are also available. 1974-01

### 3 DECORATIVE METAL DRAPERY FABRIC

A 76 per cent steel, 24 per cent polyester-blend textile incorporating type graphics for drapery and panel-wrapping. Available up to 59 in./150 cm wide, the textile can be dry-cleaned. 2345-04

### 4 EMI-SHIELDING TEXTILE

Metal-coated (nickel and nickel/silver) fabrics (wovens, nonwovens and knits), filaments and yarns, as well as finished articles, that are electrically conducting and highly flexible. Intended for electromagnetic (EMI) and radio-frequency interference (RFI) shielding. 2965-01

### 5 EMI-SHIELDING TEXTILE

A lightweight shielding system made of a nonwoven metallized material that is applied like wallpaper to permanently protect electronic equipment against electromagnetic interference in computer and industrial control rooms and laboratories. 2543-01

**PEER INC.**
ALUMTE
While strong and very durable, these sintered aluminum sheets are porous, allowing them to absorb sound effectively.

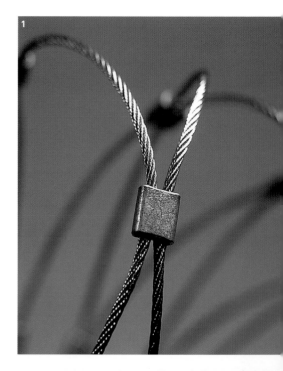

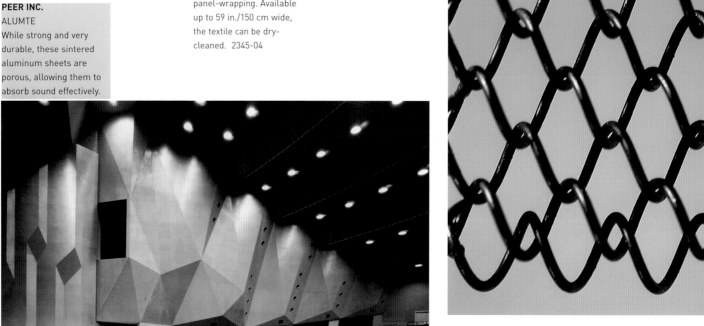

3

4

5

**1 THREE-DIMENSIONAL SPACER METAL FABRIC**
A woven polyester/nickel fabric used for low- and high-frequency electromagnetic protection. Sold in 56 in. by 100 yd/38 cm by 91 m rolls. Applications include tents, buildings and shelters.  2394-02

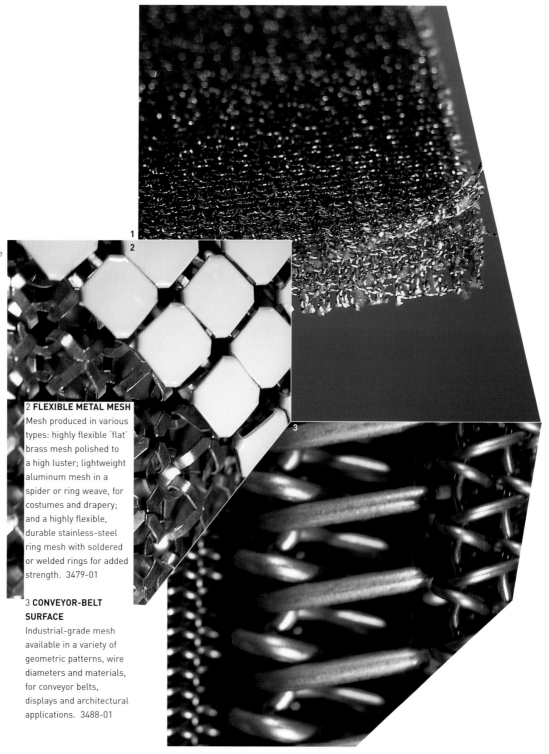

**2 FLEXIBLE METAL MESH**
Mesh produced in various types: highly flexible 'flat' brass mesh polished to a high luster; lightweight aluminum mesh in a spider or ring weave, for costumes and drapery; and a highly flexible, durable stainless-steel ring mesh with soldered or welded rings for added strength.  3479-01

**3 CONVEYOR-BELT SURFACE**
Industrial-grade mesh available in a variety of geometric patterns, wire diameters and materials, for conveyor belts, displays and architectural applications.  3488-01

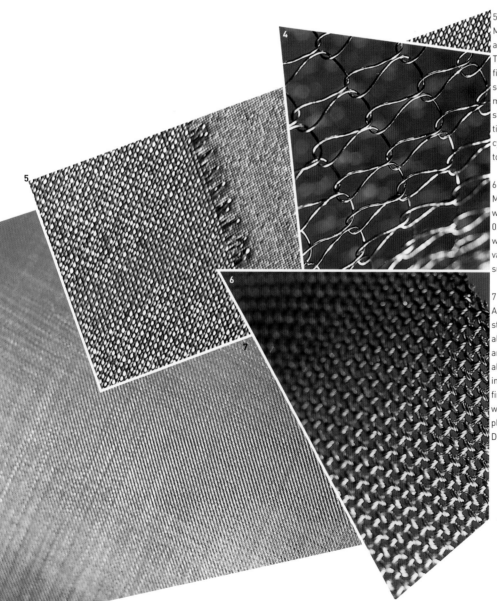

**4 KNITTED-WIRE TEXTILE**
Wire mesh knitted on conventional, circular knitting machines in stainless steel and other metals. Available in various densities and in 1 to 43 in./2.54 to 109.2 cm widths, for applications such as components for automotive exhaust systems and airbags. 4384-01

**5 FILTRATION MEMBRANE**
Metal membranes that have a uniform filtration surface. These metal-wound, sintered filters are made from seamless, continuous flat metal ribbons of an alloy of stainless steel (347SS) and titanium that are wound onto cylindrical or conical mandrels to create shapes. 4211-01

**6 WIRE CLOTH**
Meshes woven using drawn wire with diameters of 0.4 to 0.008 in./10 to 0.2 mm. The weaving process creates variations of the mesh's surface texture. 1437-01

**7 WIRE CLOTH**
Any malleable alloy – including stainless steel, titanium, aluminum, nickel, Inconel and other corrosion-resistant alloys – can be woven into this industrial wire cloth. Coarse or fine meshes are available, as well as four weaves: plain, plain Dutch, twilled and twilled Dutch. 1276-01

**05**
**METALS**
FIBERS + TEXTILES + MESH

**1 WIRE GRID**
Standard- and custom-woven wire cloth available with mesh openings from 2 to 0.0002 in./5.1 cm to 5 μ. Various types of weaves and densities, and a variety of metals including several stainless-steel alloys, copper, brass, aluminum and nickel, are available. 1489-01

**2 WOVEN-METAL PANEL**
Woven panels of acrylic-coated brass and stainless steel in a variety of patterns and textures that are scratch-, dent- and corrosion-resistant. The panels can be polished, finished and combined with different background colors to create a desired custom appearance. 1489-02

**3 DRAWN-WIRE MESH**
The weaving process creates deformations that affect the surface of this mesh. Applications include façades, wall coverings, ceilings, gates, sunshades, displays and exhibition stands, lamps, bars, racks and desks. 1575-01

**4 WIRE CLOTH**
Cloth woven in several different styles, including plain Dutch weave, plain twilled weave, broad-mesh twilled weave and reversed Dutch weave. It can be woven from more than twenty-five alloys including stainless steel, nickel-based superalloys, copper, brass and aluminum. 2005-01

**5 STAINLESS-STEEL WIRE MESH**
Custom-woven wire with mesh openings from 0.0008 to 0.6 in./20 μ to 16 mm. It is high-strength, durable, corrosion-resistant, cleanable and available in five different weaves. Applications include window-treatment panels. 2206-01

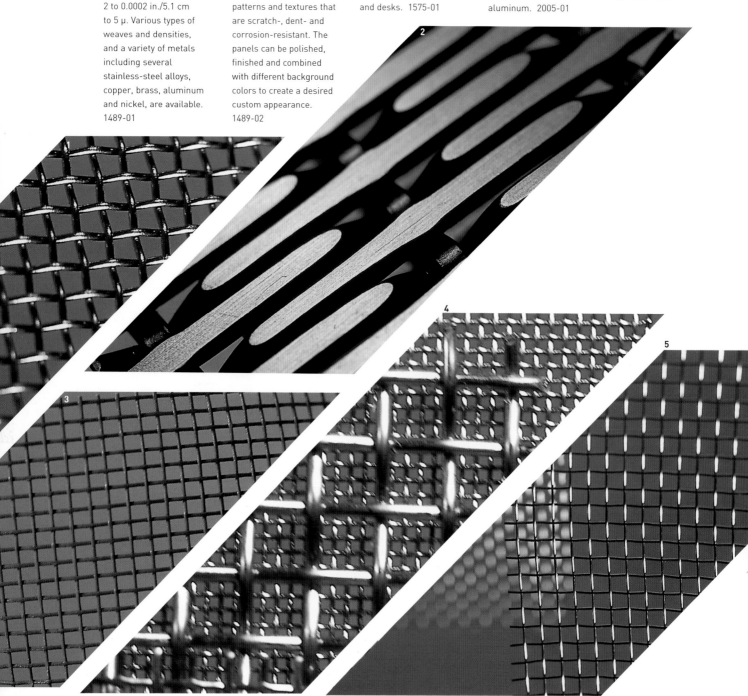

**6 ARAMID METAL FIBER**
A man-made aramidimide
fiber composed of a
Kermel®/viscose blend.
It is lightweight and
thermostable, and has
a high tear and tensile
strength. Inherently non-
flammable, the fabric will
remain physically intact on
exposure to flame.
238-301

**05**
**METALS**
FIBERS + TEXTILES + MESH

**7 SLIT METAL FOILS**
Metal and alloy foils (in
some cases plastics) with
a thickness of 0.001 to
0.005 in./0.03 to 0.13 mm),
expanded to various mesh
openings by slitting and
stretching a single roll
of very thin foil to
create gaps with exact
dimensions.  2989-01

**8 WIRE MESH**
Strong, durable welded-
wire screens that
efficiently separate fluids
and solids by using
a surface wire that is
V-shaped in profile.
This creates slots which
enlarge inwardly to
prevent particles from
lodging in and plugging
them.  3002-02

6

7

8

**CARL STAHL**
X-TEND
This non-structural open-mesh boundary is made of cold-drawn stainless steel. It is used primarily for safety and security systems, creating a lightweight physical barrier while allowing visual transparency.

**1 RIGID METAL MESH**
An architectural screening material offering containment and security with minimal visual restriction. Available in a number of finishes including copper, brass, bronze, galvanized, aluminized, flat black oxidized and painted. 3484-01

**2 WIRE MESH**
Hand-woven, stainless-steel mesh produced for animal enclosures. It has double half-hitches at every union or joint and can be made of cable with diameters ranging from 1/16 to 3/16 in./1.6 to 4.8 mm. 3484-02

1

2

**3 FLEXIBLE METAL MESH**
Highly flexible, 'flat' brass mesh polished to a high luster. The mesh can be coated with natural or metallic enamel finishes such as gold, silver, bronze and platinum. 3479-01

**4 MALLEABLE PERFORATED ALUMINUM**
A flat sheet metal punched with a trilobed pattern that has been developed for copying three-dimensional forms. It can be processed like wire grating and also shaped by hand. 3810-01

**5 SINTERED FILTRATION SURFACE**
These filters are made from seamless, continuous, flat metal ribbons of titanium and a stainless steel alloy (347SS) that are wound onto cylindrical or conical mandrels. They have a uniform filtration surface. 4211-01

**6 DRYING FILTER**
Wire cloths or meshes woven from stainless steel, bronze or FDA-grade polyester. Used primarily in paper machines and in the food industry as drying belts and filters. 4295-01

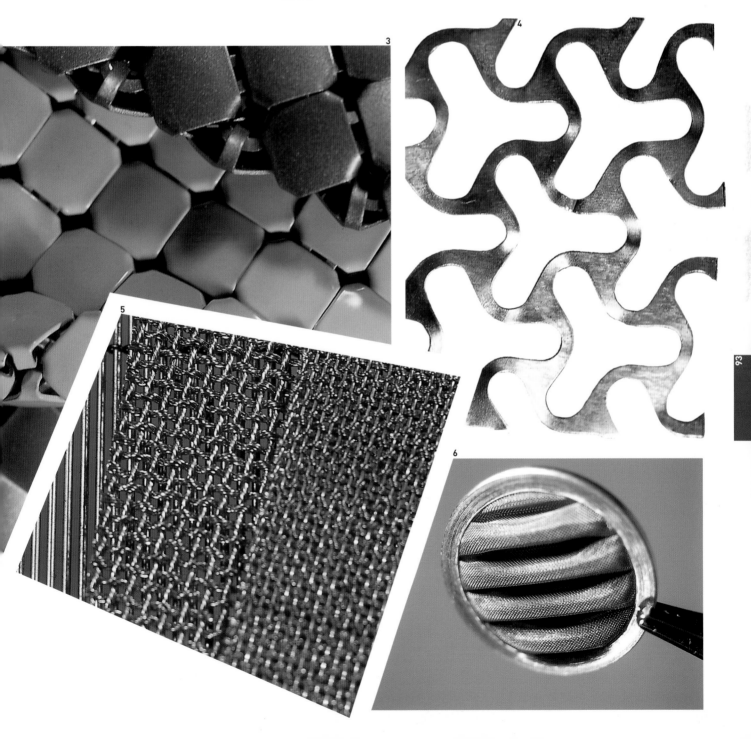

1

2

3

4

**1 STEEL FILTRATION MEMBRANE**
Three-dimensional, nonwoven metallic membranes composed of sintered, uniformly laid fine steel fibers. Due to their high porosity and dirt-holding capacity, they are often used as oil, fuel and gas filters in the polymer and chemical industry, and in airbag systems. 2699-01

**2 BEAD CHAIN**
The dumbbell-shaped connectors that link the beads together leave them completely flexible yet incapable of kinking, binding or jamming. Available in brass, several types of steel, aluminum and custom metals. 3450-01

**3 BRONZE-WIRE SUN SCREEN**
Custom-built, exterior solar screens woven from bronze wires and configured into a mesh. Made as a miniature louver system, they are designed to provide maximum outward viewing while blocking the sun's direct rays. 3644-01

**4 WOVEN-WIRE MESH**
Decorative textiles that are hand-mined, -extruded and -woven from copper, aluminum or brass. The heating and extruding processes used may cause some natural color change. 3493-01

**1 ACOUSTIC PANELING**
Sound-absorbing, durable, nonflammable, corrosion-resistant, porous aluminum material with a controlled pore size, porosity and thickness. It is fabricated of aluminum fibers sandwiched between a metal mesh and then rolled and pressed into a cohesive rigid sheet. 1049-02

**2 METAL GRILLE**
Weld-lock-constructed aluminum, brass, bronze or stainless-steel bar grilles fabricated in both custom and stock sizes and shapes. The aluminum grilles are available in various finishes such as satin- and mirror-polish and primed and anodized colors. 166-01

**3 STEEL FLOOR TILE**
Durable, textured, slip-resistant stainless-steel tiles that are 0.25 in./6 mm thick. This removable, low-maintenance floor covering can be easily fitted and installed over existing floors. 1661-01

**4 SINTERED METAL PANELING**
Sintering turns this mixture of aluminum particles and pre-alloyed metal powders into rigid but porous decorative panels that are strong, durable and sound-dampening. 1049-01

**5 MODULAR FLOORING**
A flexible floor system. The mats are constructed with frames of warp-resistant aluminum, brass or stainless steel connected by polyvinyl chloride (PVC)-coated steel wire, with inserted honeycomb rubber mats or brushstrips. 4203-01

**6 HAVER & BOECKER**
WOVEN-WIRE CLOTH
Individual steel wires are flexible by themselves. When joined together by both weaving and spot-welding, they take on new properties, becoming rigid and potentially structural.

**7 STABILIZED ALUMINUM FOAM**
Continuously cast, closed-cell foam that combines scrap and specialty aluminum alloys with a patented continuous-casting process to fabricate a lightweight panel that has the same mechanical properties in all dimensions. 3581-01

**8 STABILIZED ALUMINUM FOAM**
A lightweight, rigid, highly porous, permeable foam with a reticulated structure of open cells connected by continuous solid metal ligaments that approximate single-strand drawn wire. 129-03

**9 STABILIZED ALUMINUM FOAM**
Closed-cell foam that combines scrap and specialty aluminum alloys with a patented continuous-casting process to fabricate a lightweight aluminum panel that has the same mechanical properties in all dimensions. 3581-01

**10 STABILIZED ALUMINUM FOAM**
Lightweight and energy-absorbent foam with high stiffness and low thermal conductivity. A surface skin of aluminum encloses the cellular structure, of which air makes up 80 per cent of the total volume. 4076-01

**11 CAST ALUMINUM FOAM**
Low-density foamed aluminum with a porous cellular structure made from either a powder of wrought or cast aluminum alloys and a foam agent that is then processed into a compact precursor material before it is foamed in a closed mold. 1495-01

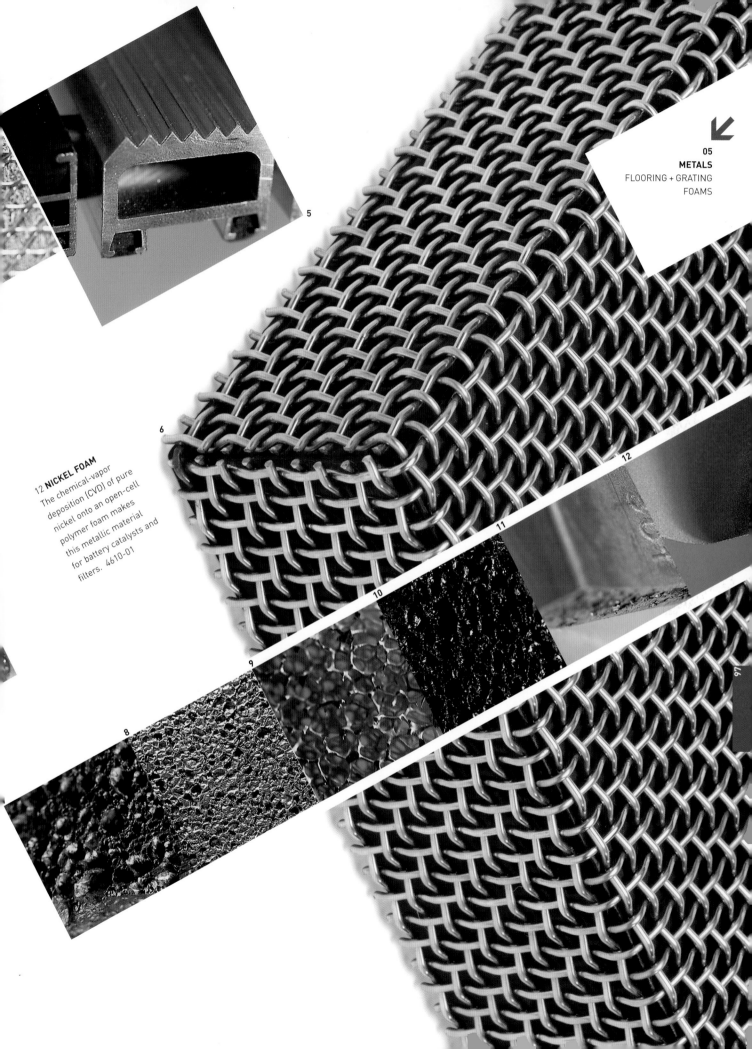

5

6

12 **NICKEL FOAM**
The chemical-vapor
deposition (CVD) of pure
nickel onto an open-cell
polymer foam makes
this metallic material
for battery catalysts and
filters. 4610-01

12

11

10

9

8

1+2 **ALUMINUM HONEYCOMB PANELING** Translucent panels with a strong, lightweight polycarbonate or polyethylene honeycomb core (honeycomb cell size 3/8 in./9.6 mm). The outer surfaces are low-maintenance, weather- and scratch-resistant fiberglass-reinforced acrylic facings in clear or colored.  3572-01

### 3 ENERGY-ABSORBENT ALUMINUM SHEET

A tile or sheet that is lightweight, low-cost and recyclable. This sheet may be easily manufactured by thermoforming and cold-pressing aluminum alloys, polycarbonate, acrylo-nitrile butadiene styrene and polypropylene, among other materials. 4564-01

### 4 CELLULAR-CORE LIGHTWEIGHT PANEL

Strong, rigid, lightweight, weatherproof and insulating laminated panels. They are composed of two face-sheets of aluminum (custom face-sheets are available) and a homogenous cellular core material that is free from inclusions. 1433-05

### 5 HONEYCOMB PANELS

Two types of lightweight, flat, honeycomb-metal wall panels. Designed to be installed as a dry-set system that allows the components to be sealed on the back, thereby eliminating caulk joints or black gaskets. 795-01

### PANELITE LAMINATES
PANELITE

Translucent honeycomb-core panels are used as interior curtain-walls for the entrance to facilities at a New York nightclub. The panels have an aluminum or polycarbonate honeycomb core sandwiched between sheets of fiberglass-reinforced resin.

**1+2 HIGH-PRESSURE LAMINATES**
A series of decorative metallic laminates with three-dimensional patterns and designs. Offered in a variety of metals: aluminum and anodized aluminum, brass, copper, bronze, brass-plated copper and chrome-plated brass. Available with or without a phenolic backer.  417-01

**3 LIGHTWEIGHT, RIGID EXTERIOR PANELING**
A composite building panel that is resistant to shock, pressure, bending and breaking. It is composed of a polyethylene core faced with two color-coated aluminum sheets (each 0.02 in./0.5 mm thick) that are bonded mechanically.  1615-01

**4 METAL-POLYSTYRENE COMPOSITE SHEET**
Sheets composed of a highly impact-resistant polystyrene base layer and either a metal-coated, polyester-film or printed surface.  4347-01

**5 TITANIUM-ZINC CLADDING**
This composite material consists of two sheets of titanium-zinc alloy that are permanently bonded to an extruded thermoplastic core material. The total thickness of the material is 0.2 in./4 mm.  4797-01

**6 LIGHTWEIGHT, RIGID EXTERIOR PANEL**
A corrosion-resistant aluminum laminate composed of two sheets of aluminum 0.02 in./0.05 cm gauge with a proprietary polyethylene core that can be processed with standard wood- and metalworking tools.  1627-03

**7 CONSTRAINED-LAYER VIBRATION-ABSORBING SHEET**
A family of sound- and noise-damping materials that utilize a core of viscoelastic thermoplastic sandwiched between two metal skins. These materials dampen vibration over a wide range of temperatures.  4371-01

**8 ALUMINUM HIGH-PRESSURE LAMINATE**
Melamine resin, decorative paper and core paper are fused with aluminum to create a layered sheet. The process of layering aluminum creates extensive structural stability. Details include lipping and curved and milled edging.  4834-01

**9 FLEXIBLE MAGNET**
Material available in a wide assortment of sizes, shapes and thicknesses, composed of strontium or barium ferrite suspended in sulfur-cured nitrile rubber. The magnet can be cut, punched and machined without losing its magnetic strength or structural integrity.  184-01

**10 MAGNETORHEOLOGICAL FLUID**
Composed of micron-sized gray magnetic particles suspended in water, oil or silicone, this fluid responds in milliseconds to a magnetic field by hardening and softens when the field is removed.  3680-01

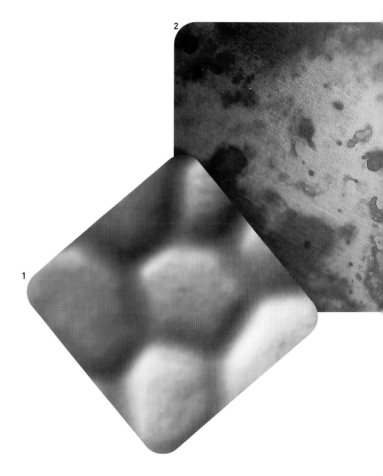

2

1

**KME**
AMSTERDAM FORUM
Copper naturally patinates green over time. The cladding on this building in Amsterdam uses copper that has been treated with a special manufacturing process that causes the copper to patinate lighter and darker shades of brown.

**MOZ DESIGNS INC.**
ALUMINUM PANEL
The three-dimensional
effects on these
aluminum sheets are
created by selective
grinding. The metal is
then anodized to harden
and color the surface
while simultaneously
enhancing the three-
dimensional appearance.

**1 SCREEN-PRINTED ALUMINUM CLADDING**
Textured, coated metal sheets that are suitable for exterior applications. A 100 per cent aluminum sheet coated using a polymeric film approximately 0.0004 to 0.0005 in./10 to 12 µ thick applied by rotational screen-printing. A wide variety of colors and designs are available. 1475-02

**2 FLEXIBLE POLYMER-COATED SHEET**
Decorative embossed, pressed, polished sheets. A 100 per cent polyvinyl chloride (PVC) polished, clear top sheet is laminated onto various aluminum-foil inner layers and then reverse-embossed to give an illusion of depth. 2604-11

**3 ARCHITECTURAL ALUMINUM PROFILE**
Corrugated, flat and curved profiles that can be perforated, drilled, milled and anodized. Virtually any corrugation profile may be produced, including variations in the frequency and amplitude of the corrugation. 108-01

**4 METAL-POLYMER COMPOSITE**
Molded tile for interior applications. A single-step molding incorporates metal, filler and a resin matrix to produce composite tiles with the appearance of solid cast metals. Copper, brass, nickel silver, bronze and gunmetal colors are available. 181-01

**5 PERFORATED METAL SHEET**
Aluminum sheets (0.04 to 0.12 in./1 to 3 mm thick) perforated in a variety of patterns, including round holes in a wide range of diameters and patterns; hexagonal holes, which are the strongest shape, in square pitch and staggered patterns; and square holes. 3681-01

1

2

3

4

5

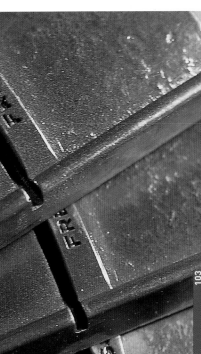

## 4 SHAPE-MEMORY ALLOY ACTUATOR

Electronic actuators for use as silent, high-precision motor replacements. Nitinol-alloy wires, which recover their original shape after deformation, contract when an electric current is passed through them. This allows for constant force/torque and very fine control of small movements. 4957-01

## 5 METAL INJECTION-MOLDING PROCESS

Injection-molding used to make small, complex metal parts. The process combines computer-aided design (CAD) with injection-molding technology in which metal powders (stainless steel, low-alloy steels) mixed with plastic binders are injected into molds. 4048-01

## 6 SHAPE-MEMORY ALLOY

These strong, flexible, kink-resistant alloys, composed primarily of nickel and titanium in addition to other metals such as copper or niobium, can recover their original shape after deformation. 4307-01

## 1 INTERIOR PANELING

Decorative aluminum paneling for vertical interior surfaces. Available perforated, it can be manufactured in sixteen colors (plus custom colors) and ten patterns. Stainless-steel versions are also available as a custom option. 3648-01

## 2 RIGID PATTERNED METAL SHEET

Thin, hexagonally structured sheet metal with a high strength-to-weight ratio. The regular convexity improves heat transfer while creating effective light-scattering and good acoustic properties. 2045-01

## 3 PERFORATED METAL SHEET

Coils, sheets or blanks up to 60 in./152 cm wide and 0.008 to 0.13 in./0.02 to 0.32 cm thick perforated with round, square, hexagonal or slotted holes in metals including steel, stainless steel, aluminum, tin, coated steel, brass and copper. 1482-01

**7 BALL CHAIN
MANUFACTURING CO.
INC.**
SHIMMERSCREEN
The serpentine curtain
hung from the ceiling is
made of strands of small
nickel-plated hollow
spheres. It undulates
around the light fixtures,
creating shadows and
moving with air currents
in the room.

7

1 **SPINNEYBECK**
WALL PANEL
The warm and pliable
leather of this wall
surface is covered
with a technically
advanced breathable,
protective, water-
resistant coating.

2+3 **ABET LAMINATI**
MEG
A digitally printed
high-pressure
laminate surface has
been developed
especially for exterior
use (as in these
curtain wall tiles),
as has a decorative
high-pressure
laminate that makes
an ideal surface for
high-traffic zones
because of its
durability. Very few
laminated surfaces
cannot be used
outdoors. While the
outer surfaces of
these laminates is a
polymer, they are
predominantly made
of paper.

The category of 'naturals' covers all
materials that come directly from a plant
or animal source. These include the
obvious candidates such as wood, cotton
or wool yarns, stone and animal hides
that are not altered in form before use,
and the natural-material derivatives such
as paper from wood-derived cellulose,
foams from soy precursors and starch-
based plastics.

Although it may not be possible to develop
a new wood in the same way that new
polymers are synthesized, there are
nevertheless innovations in the area of
natural materials that have utilized
woods, agricultural by-products and
fibers even in high-performance
applications. These developments include
new shaping processes for wood and
wood veneers, alternative uses for fibers
such as flax, hemp, flax jute, kenaf and
sisal, and the creation of biopolymers
from corn, potatoes and soy.

Recent advances in veneers have
broadened noticeably the areas in
which natural material surfaces may
be utilized. The incorporation of resins
into the veneers have rendered them
as thin and flexible as paper, allowing
their use as alternatives to textiles in
padded upholstery (page 132/2), as
light-transmitting panels in lighting
applications and even as apparel accents.
Further utilizing this flexibility of thin
sheets of veneer, compound curves may
now be shaped from successive layers,
creating sculptural plywood forms
that redefine the limitations of wood
(page 135/6).

There has been a considerable increase in
the use of wood fibers as strengthening
additives to plastics for automotive
applications, as well as significant use
of long agricultural fibers such as sisal,
hemp, kenaf and others that offer some
advantages over more traditional glass
and synthetic fibers. Fiberboard products
have made use of waste straw (page
113/7), sugar cane, recycled paper
products and sunflower-seed husks
(page 115/7) as alternatives to wood
fibers. Advances in formaldehyde-free
and soy-based resin binders have allowed
manufacturers of such fiberboard to offer
more 'environmentally friendly' solutions.

It is now possible to create polymeric
materials from annually renewable
natural resources such as potato starch,

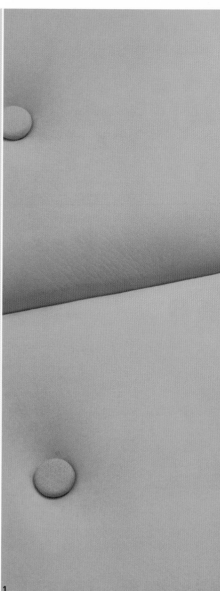

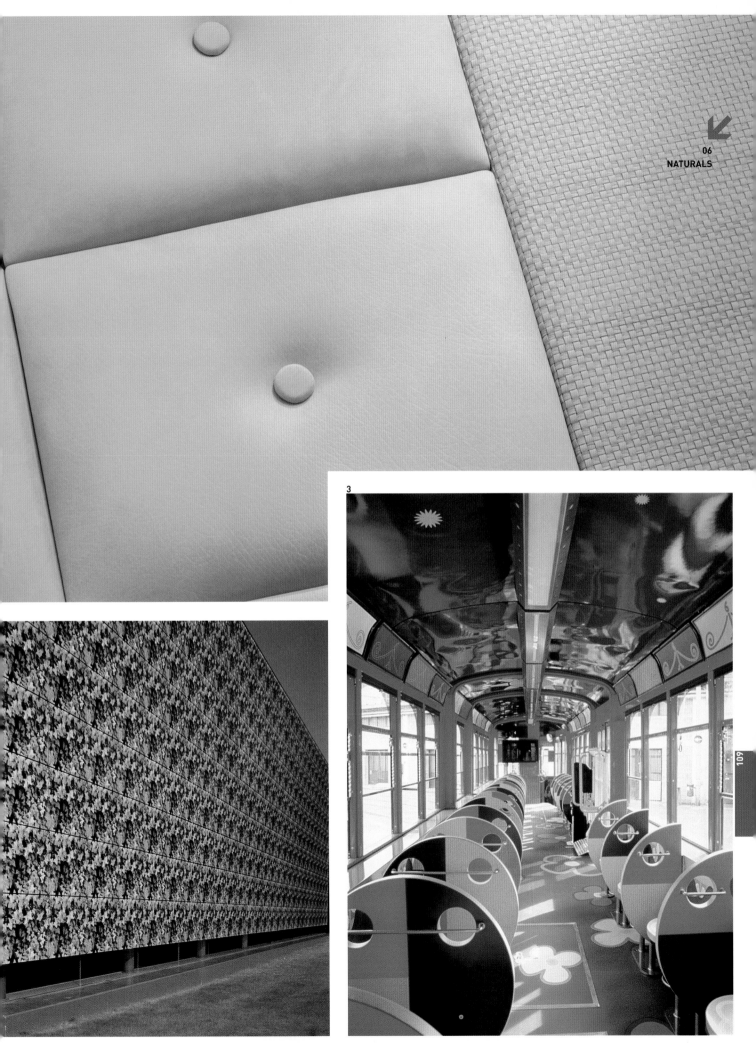

3

1

1 **MATERIALISE (FREEDOM OF CREATION)**
LILY
The plastic petals of this translucent lamp are created using a rapid prototyping process. Successive two-dimensional layers are written one on top of another, very much like an ink jet-printed piece of paper. There are no limits to the design and shape of objects created using this process.

2 **MIO CULTURE**
FIBRID
Paper you can sit on, this 100 per cent recycled pulped paper is made from newsprint and cardboard. Though rigid and lightweight, it is not waterproof.

3

corn and soy. Of these 'natural' or biopolymers, those formed from corn sugars have shown the most promising results, producing polyester-like resins by synthesizing polylactide (PLA) to form the biopolymer. The resin can be extruded (into fibers), thermoformed and blow-molded, and after use it can be recycled into monomers and polymers. Current forms include fibers, nonwovens (textiles that are neither woven or knitted but made in a manner similar to felt manufacture), films, packaging, coatings and wraps. The fibers readily accept dyes and fillers, can be co-polymerized with other materials to further enhance performance, and are most comparable in properties to polyester and polystyrene.

Polyurethane foams (page 142/1) and resins have also been produced via a natural route, by synthesizing soy proteins. The end product is chemically equivalent to polyurethane and as such has the same properties as that which comes from the petrochemical-based polymer, but it is from a sustainable resource. Other natural-based plastic alternatives include injection-molded parts created using lignin, a dark, sticky waste product left after the extraction of cellulose from wood (page 142/4). These molded parts still retain the smell of wood and are an appealing alternative to plastic resins.

The increased use of wood, agricultural fibers and other by-products that would previously have been discarded as waste

2

bodes well for our appreciation of the range of natural materials. The growth of biopolymer applications used in preference to petrochemical polymers perhaps heralds a breakthrough in attempts to reduce our reliance on petrochemicals. Above all, it is clear that despite an obvious desire for the synthetic, innovations in naturals are allowing us to satisfy our need for the use of these materials in more and more diverse applications.

**3 AVEDA**
URUKU LIPSTICK
CASE
A reusable lipstick
case composed of
postconsumer
recycled (PCR)
polypropylene and
30 per cent natural
flax fibers.

**4  REHOLZ**
GUBI
The oak-plywood
shell of these chairs
use non-
perpendicular layers
of plywood, allowing
the formation of
complex compound
curves. The shapes
are created under
high pressure and
elevated
temperatures inside
a steel mold.

**06**
**NATURALS**
BAMBOO

**1 BAMBOO VENEER**
Veneers made of
10 per cent bamboo
sheets on plywood or
cotton cloth. Ideal for
interior applications, they
are available in either
natural-unfinished or
carbonized-unfinished
surface colors.  5015-01

**2 BAMBOO FLOORING**
A hard material with good
dimensional stability,
constructed of laminated
bamboo planks or
centered tongue-and-
groove, and resistant
to scuffing, staining,
moisture, discoloration
and mildew.  2016-01

**3 BAMBOO FLOORING**
Bamboo for woodworking
and flooring that is
ecologically sustainable.
It is constructed of
98 per cent precision-
milled bamboo strips
laminated with UV-cured
PVAC formaldehyde-free
glue, and with the surface
planed to dimension.
4235-01

**4 BAMBOO FLOORING**
This tongue-and-groove,
laminated flooring is
available as panels or
boards, finished or
unfinished, in a natural
color or carbonized. It also
can be colored. Bamboo is
as hard as hard maple and
more stable than red oak.
2017-01

**5 BAMBOO FLOORING**
Hard, stable, easily
installed flooring
constructed from layers
of bamboo strips that
are laminated under high
pressure. Harvested
and manufactured in its
native Chinese habitat,
the material is easily
replenished and therefore
more sustainable than
hardwood.  56-01

**6 BAMBOO WINDOW
SCREEN**
Hand-made window
screens with fine strips
of bamboo in the weft and
natural monofilament in
the warp. The screens are
coated with a natural tree
lacquer.  4815-01

1

3

4

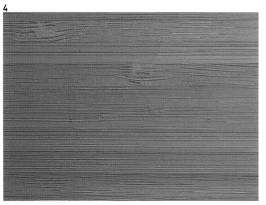

5

6

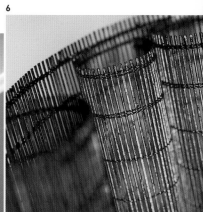

**7 FORMALDEHYDE-FREE MEDIUM-DENSITY FIBERBOARD**
A golden-brown, engineered composite panel with good water resistance and screw-holding ability. It is made from natural fibers found in crop residue and bonded with proprietary formaldehyde-free binders. 4-03

**8 NATURAL FIBER-POLYPROPYLENE COMPOSITE PANEL**
Sheets made from recycled post-industrial plastics for the building and marine industries. Composed primarily of polypropylene and cellulose fibers including sawdust, coffee and kenaf. The sheet can be covered on one or two sides with Trevira®. 4704-01

**9 POLYMER-WOOD COMPOSITE DECKING**
A moisture-resistant decking material composed of steam-treated and pre-cleaned cedar fibers encapsulated in a blend of low- and high-density polyethylene. Unlike wood, this material does not splinter or crack, nor does it require chemical dipping or preservative treatments. 95-01

**10 WOOD-POLYMER COMPOSITE DECKING**
A decking material made primarily from equal parts of reclaimed hardwood sawdust and reclaimed and/or recycled polyethylene plastic. Stable and workable, it provides traction and contains no preservatives or toxic chemicals. 81-01

7

8

9

10

**1 BIOCOMPOSITE PANELING**
A biocomposite made of 40 per cent recycled newsprint-paper products, 40 per cent soy-based resin system and 20 per cent color additives and other materials. These components are bonded with proprietary formaldehyde-free binders.  0003-01

**2 POST-CONSUMER-WASTE CONSTRUCTION BOARD**
Construction boards for office furniture that are fully recyclable, moisture-resistant, noise-dampening and -absorbing, easily reshaped when heated and suitable for most types of finishing techniques including laminating, veneering and lacquering.  1444-01

**3 INSULATING CONCRETE**
Concrete forms that are strong, sound-absorbing and thermal insulating. They are also termite- and pest-resistant; weather-, rot- and decay-proof; have a four-hour fire rating; and are sufficiently porous to allow the slow transference of air. 4405-01

**4 RELIEF-CARVED FIBERBOARD PANEL**
Panels carved with a variety of decorative patterns. The carving is done with many different materials including medium-density fiberboard (MDF), laminated veneers, solid wood and materials composed of mineral fibers.  4184-01

## 5 GYPSUM WALLBOARD

A wallboard composed of a fiberglass-mesh scrim embedded in a high-density layer of gypsum and cellulose fibers, followed by a thin, lightweight core and another high-density layer of gypsum mixed with fiber. It is abuse- and fire-resistant. 1155-01

## 6 STRUCTURAL COMPOSITE LUMBER

Flooring and decorative surfaces made of structural composite lumber (SCL). This shrink-resistant, formaldehyde-free engineered wood surface is manufactured of strands that are up to 12 in./30.5 cm long. 4489-01

## 7 SUNFLOWER-SEED FIBERBOARD

A bio-based composite material with the appearance of traditional burled wood. It is produced from 84 per cent sunflower hulls from North and South Dakota and bonded with a formaldehyde-free, 10 per cent diphenylmethane diisocyanate (MDI) binder. 4-02

## 8 CORRUGATED FIBERBOARD

Corrugated (B flute), standard medium-density fiberboard (MDF). A specially designed machine cuts and removes longitudinal strips from its surface. 3964-02

1

2

3

4

**1 WOOD-CEMENT COMPOSITE ACOUSTIC PANEL**
A wall-panel system wrapped with an abuse-resistant fabric that is strong, tough and durable, and controls noise in open, closed or mixed-use spaces. The panels are fabricated of aspen-wood fibers plus a hydraulic cement binder. 493-01

**2 WOOD-CEMENT COMPOSITE ACOUSTIC PANEL**
Structurally strong, lightweight panels that have a Class A/I flame-spread rating and a decorative textured finish. They are composed of aspen-wood fibers bonded with a hydraulic cement binder. 493-02

**3 STRAW-BASED PARTICLEBOARD**
Industrial-grade particleboard made from agricultural residue. Straw-based particleboard is bonded with a proprietary non-toxic, emission-free binding agent. 1087-01

**4 STRUCTURAL COMPOSITE LUMBER**
Laminating yellow poplar, western hemlock, Douglas fir or southern pine strand lumber creates these large, strong and stiff cross-section headers, beams and columns designed to support floor and roofing systems. 1392-02

5

5 **THROUGH-THICKNESS COLORED FIBERBOARD**
Pigmented fiberboard derived from pine fibers extracted from Portuguese forests. Organic pigments are added to wood fiber and combined with a melamine-urea-formaldehyde binder to produce a fiberboard that is colored throughout.
4850-01

6 **HEAT-INSULATING PANEL**
Cellulose-based fiberboard that contains at least 65 per cent PCM (phase-change material) paraffin. (PCM's are highly crystalline paraffin/wax materials that have been added to fiberboard).
4756-02

7 **CHICKEN FEATHER-SOY COMPOSITE**
Hollow keratin fibers (from chicken feathers) are combined with chemically modified soybean oil and then compressed to form these composite boards, which are suitable for use as printed circuit boards.
4863-01

8 **STRAW-BASED MEDIUM-DENSITY FIBERBOARD**
Strong, machinable panels made from straw fibers (the residue of wheat) and non-toxic polyurethane resins. The panels have good moisture-swell, elasticity, internal-bond and density properties without harmful emissions. 3657-01

6

7

8

117

**06**
**NATURALS**
HONEYCOMB
MATERIALS

2 **HONEYCOMB-CORE COMPOSITE BOARD**
Honeycomb composites for lightweight structural paneling. This honeycomb is a kraft-paper product fabricated to form a continuous series of triangular cells, similar in appearance to the cross-section view of corrugated board. 127-02

4 **PAPERBOARD HONEYCOMB SANDWICH CORE**
Honeycomb core panels from cellulose. Honeycomb structures are sandwiched between cardboard sheets to produce lightweight, impact-resistant board. A range of compression strengths, surface tolerances and minimum moisture absorbencies are available. 1438-01

1 **PAPERBOARD-COATED FIBERBOARD**
A strong, lightweight sandwich panel. It is composed of a highly dense corrugated core of engineered wood fibers sandwiched between twin faces of a formaldehyde-free, water-resistant polymer-impregnated paperboard. 3990-01

3 **PAPERBOARD HONEYCOMB SANDWICH CORE**
A lightweight honeycomb core made of kraft paper formed into a series of triangular cells. These are bonded on both sides to a variety of facing materials. The resulting sandwich panels have low densities but good load-carrying capacity. 127-01

5 **CORRUGATED PLYWOOD PANEL**
Aramid paper sandwiched between various wood veneers and bonded with formaldehyde-free glue. The panels are fire-resistant and have a very high tensile strength. 1483-01

3

1

2

6 **PAPERBOARD HONEYCOMB SANDWICH CORE**
Strong, lightweight honeycomb cores composed of a series of kraft-paper ribbons glued together to form hexagonal cells that are faced on both sides. Continuous cores can be laminated as slices and as expanded panels.  2476-01

7 **HONEYCOMB-FOAM SANDWICH PANEL**
Sandwich panels that are custom made of honeycomb and foam cores laminated with a variety of face materials, depending on the application. Honeycomb core materials include aluminum, treated kraft paper, foam, foam-filled kraft paper, aramid fiber and thermoplastic.  102-01

5

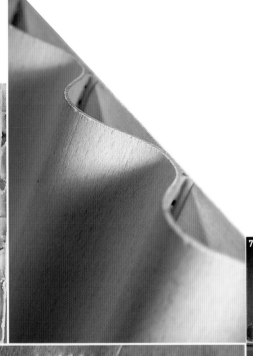

4

6

7

**1 LAMINATED WOVEN-WOOD VENEER**

Woven-wood veneers fabricated from 1.6 to 2.4 in./40 to 60 mm strips available in 120 different varieties of wood that are laminated onto a mahogany substrate; used for furniture and interior architectural applications. 4184-02

**2 SOUND-ABSORBING PANELING**

Hard-surfaced paneling for ceilings, walls and partitions. Manufactured from MDF with a wood-grain pattern or solid-color laminate surface. 4533-01

**3 RECYCLED-WOOD PARTICLEBOARD**

Industrial particleboard that has a flat and smooth surface. It is made from 100 per cent-certified post-manufactured wood and reaches the Green Cross Certification standards, as well as US Housing and Urban Development standards for 0.3 ppm urea-formaldehyde emission. 4223-01

**4 HIGH-PRESSURE LAMINATE FLOORING**

A tough, durable flooring made of high-density fiberboard (HDF) which has high-pressure laminate bonded to both sides of its kraft core, thus making it resistant to indentation by heels and heavy objects. 91-04

**5 ACOUSTIC-ABSORPTION WOOD PANELING**

A natural wood-veneered panel. The wood is laminated to an acoustical core. The patented perforations and grooves run either horizontally or vertically – the direction of the grain and groove must match. 1451-01

1

2

3

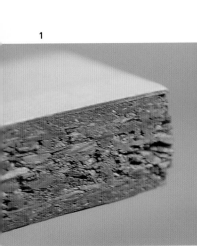

4

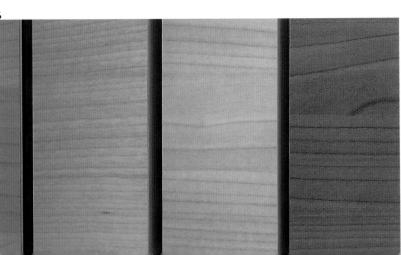

#### 6 BIRCH-VENEER WALL PANEL

Panels for interior applications. Veneers on MDF or plywood panels are pressed with repeating patterns or inlaid with brass or aluminum shapes. Standard panel sizes are 23 by 47 in./ 60 by 120 cm or 12 by 60 in./30 by 150 cm, and 0.5 in./13 mm thick. 4553-01

#### 7 MODULAR WALL TILE

A system of individual square panels that have a tongue-and-groove joint on all four beveled edges, which fit together to form a grid pattern. The panels are available in standard 2 by 2 ft/0.6 by 0.6 m sizes, with fifteen other square and rectangular sizes available. 9-01

#### 8 CURVABLE EMBOSSED PANEL

Panels which can be fabricated into curved and other custom shapes. Core materials include medium-density fiberboard (MDF), particleboard and solid wood. Surface finishes include laminated aluminum, lacquered paper and veneer. 2239-01

5

#### 9 CEMENT-WOOD COMPOSITE PANEL

Pre-cast textured slabs composed of extra-long, fine wood fibers that have been chemically processed and pressure-bonded with waterproof Portland cement. They are lightweight, strong and fire-retardant with insulating and acoustical properties. 62-01

6

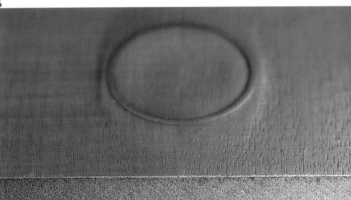

7

8

9

**1 WOVEN LEATHER
FLOORING**
Woven leather loomed
58 in./14.3 cm wide that
can be customized for
color, weave, width of
thong, grain or suede,
or both. It is combined
with another fabric for
furniture, walls and rugs.
21-04

**2 DECORATIVE TEXTURED
LEATHER**
100 per cent horsehide is
laser-cut into decorative
patterns on the hair side.
Four laser-cut patterns
are available, each with a
set single color. 71-02

**3 LEATHER FLOOR TILE**
This 100 per cent full-
grain, aniline-dyed
European cowhide and
calfskin is cut from the
center portion of the hide,
where the fibers are the
tightest. Specifically
intended for floors, it can
also be used for vertical
surfaces. 47-01

**4 LEATHER TILE**
Tiles made using the
same technology that
makes shoe soles.
They are 100 per cent
vegetable-tanned,
uncorrected leather
and contain no additives
or fillers. 4696-01

1
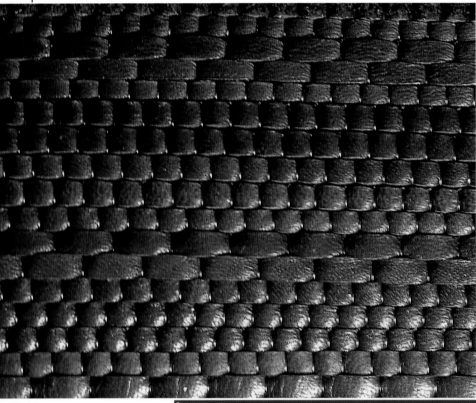

2

3

5 **TEXTURED LEATHER**
Laminated leather with a textured surface effect. High-tech leather with thin (less than 0.006 in./ 0.15 mm) polyurethane finishes laminated onto the outer surface. The material features high water-repellency on the PU-finished side. 4567-01

6 **WOVEN LEATHER TEXTILES**
Textiles woven on a fine nylon-thread warp in combinations of leather, suede and linen. A patented machine cuts the hides in a circular fashion, thereby enabling the weaver to work with very long, continuous 'threads'. 3297-01

7 **STINGRAY SKIN**
The preserved, dried skin of a particular type of stingray. It can be laminated and is used for interior decoration and design. 21-01

1 **ELECTRICALLY INSULATING PAPER**
Vulcanized fiber from a cellulose base. High-purity cellulose papers are chemically laminated to create electrically insulating sheet material. 3426-01

2 **HOLOGRAPHIC PAPER**
Metallized security paper that integrates covert security holograms into the paper itself can accommodate two- and three-dimensional images and has the flexible converting capabilities and printing properties of conventional paper. 3617-03

3 **WOVEN-PAPER FABRIC**
Durable woven fabrics for upholstery and wall coverings made of paper fibers woven with cellulose-based fibers such as linen, cotton and jute; available in a variety of designs. The fabrics are easily cleaned and come in nine standard colors. 3297-02

4 **RECYCLED MONEY PAPER**
Printable paper made from money. Acid-free papers made using 30 per cent post-consumer waste from US currency that would otherwise be destroyed, thus protecting the environment by decreasing the production of carbon dioxide $(CO_2)$ and ash created by the burning process. 4174-04

**5 MECHANICALLY
FASTENING CORRUGATED
CARDBOARD**
Standard corrugated
cardboard (flexible B flute;
49 flutes/ft and 1/8 in./
0.3 cm thick) with strips
that become engaged for
mechanical fastening. It
is processed on a specially
designed machine that
cuts and removes
longitudinal strips of paper
from a roll.  3964-01

**6 PRESSED CORRUGATED
BOARD**
Lightweight board made
of 100 per cent wood pulp
hot-pressed to form the
corrugation without the
use of an adhesive;
available in sheets up to
124 by 204 in./3.2 by 5.2 m
in three thicknesses, each
with different corrugation
frequency and amplitude.
106-01

**7 LAMINATED
PAPERBOARD TUBE**
Laminating together
multiple layers of
heavyweight fiberboard
forms these strong tubes.
They can be produced
in a variety of sizes and
customizable shapes.
3854-01

2 **RECYCLED PULP PACKING**

Packaging material of precision-molded pulp made from recycled and biodegradable paper. Used mainly for protecting products in transit, it has a smooth surface on both sides and can be used to display products in a retail setting. 1586-01

3 **HAND-DIPPED PAPER LACE**

Hand-made paper and paper lace traditionally produced by dipping a fine screen into liquified plant-fiber pulp. 4620-01

1 **HOLOGRAPHIC PAPER**

The surface of this metallized, micro-embossed paper diffracts light like a hologram. However, it has all the flexible converting capabilities and printing properties of conventional paper. 3617-01

4 **OVEN-SAFE PAPERBOARD TRAY**

Microwavable/oven-safe pressed-paperboard trays that reduce cooking hot spots and can be used at temperatures up to 400ºF/204ºC. They also can be stored in the freezer or refridgerator without cracking or denting. 1565-01

## 5 RECYCLED-NEWSPRINT WALLBOARD

A resilient, acoustically insulating, durable building board made from 100 per cent recycled newsprint. It has twice the insulation value of wood but has no splinters, voids or knots. It is paintable and tackable. 135-01

## 6 *WASHI* PAPER

Natural, hand-made Japanese paper (8.9 by 6.8 ft/2.7 by 2.1 m sheets) available in designs or custom-made. For interior panels in doors, partitions and lighting. It can be enclosed in glass. 179-01

## 7 NATURAL VELLUM

Translucent vellum papers made from pure cellulose fiber. No resins are added to increase their translucency, nor is acid or chlorine. They are recyclable and biodegradable, and can be printed and finished using a wide range of processes. 3410-01

## 8 HAND-MADE PAPER

These papers are made primarily from the pulp of the salago, a shrub that grows wild in the Philippines. The other natural fibers added for texture and design include abaca threads, rice hulls and stalks, as well as banana bark and stem fibers. 3271-01

**1 G.T. DESIGN**
MARRONE TERRA SCURA
AND ROSSO IBISCUS
This evocative design is
woven by Indian craftsmen
out of coir, the coarse
structural fiber of the
coconut fruit, which is
resistant to fungi and
mold, and is biodegradable
and anti-static. The use
of coir for floor coverings
and other domestic
applications increases the
resource utility of the
coconut fruit beyond its
use as a food source.

**3 NATURAL-FIBER**
**PERFORATED SHEET**
Perforated natural-fiber
and polyurethane sheets,
which are three-
dimensionally formed in
a variety of shapes that
are lightweight, stable,
permeable and flexible
(can be varied from elastic
to stiff). 1886-05

**4 MICA-PLATELET**
**PIGMENT**
Pearlescent and iridescent
pigments composed of
mica platelets coated with
titanium dioxide and/or
iron oxide. The pigments
show multiple color effects
and can also show two
different colors when
viewed at two different
angles in coating systems.
2040-01

**5 JUTE-HAIR CARPET**
**CUSHION**
Carpet cushion made of
90 per cent jute and
10 per cent hair. Waffled
latex rubber coated on
both sides of the jute
fiber provides a clean,
skid-proof surface.
For commercial and
residential use. 1982-01

**2 CELLULOSE FIBER**
A cellulose-based fiber
that is soft, smooth,
strong and 100 per cent
biodegradable, and can be
blended with other fibers.
Available in both medical
and textile grades.
1935-01

1

2
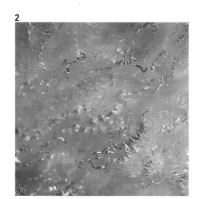

3
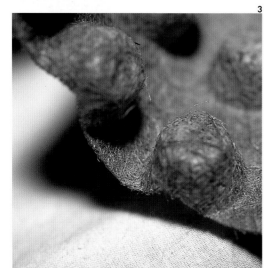

4
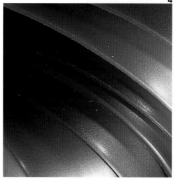

5
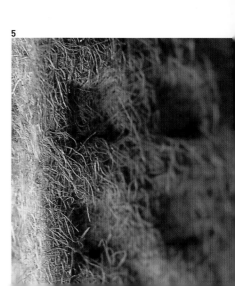

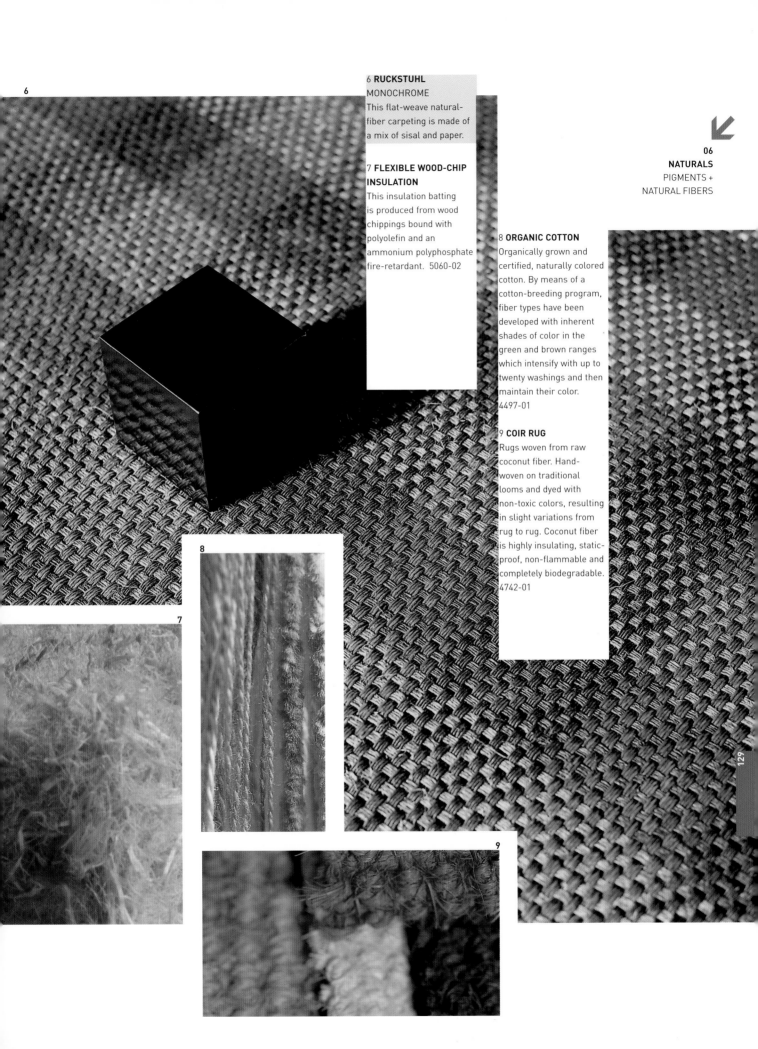

6

**6 RUCKSTUHL MONOCHROME**
This flat-weave natural-fiber carpeting is made of a mix of sisal and paper.

**7 FLEXIBLE WOOD-CHIP INSULATION**
This insulation batting is produced from wood chippings bound with polyolefin and an ammonium polyphosphate fire-retardant. 5060-02

**8 ORGANIC COTTON**
Organically grown and certified, naturally colored cotton. By means of a cotton-breeding program, fiber types have been developed with inherent shades of color in the green and brown ranges which intensify with up to twenty washings and then maintain their color. 4497-01

**9 COIR RUG**
Rugs woven from raw coconut fiber. Hand-woven on traditional looms and dyed with non-toxic colors, resulting in slight variations from rug to rug. Coconut fiber is highly insulating, static-proof, non-flammable and completely biodegradable. 4742-01

7

8

9

1 **HOG-HAIR MATTING**
Fabricated from a blend of
hog's hair and plant fibers
and coated with a latex
binder, this custom-
molded sheet packaging
material is UV- and
fungus-resistant and
flame-retardant. 100-03

2 **SEAWEED FIBERS**
The fiber in this apparel
textile incorporates
seaweed cellulose,
cellulose from other
plant material and silver.
4855-02

3 **COIR-FIBER
INSULATION MAT**
A firm, strong and
durable insulator pad
for mattresses made of
coir (coconut) fibers that
are needle-punched and
treated with proprietary
binding agents. Available
in five different grades.
154-01

4 **FLAX INSULATION**
A natural insulating
material that is easy to
install. It is both thermally
and acoustically insulating,
absorbs moisture at high
humidity and releases it
when the humidity drops,
and allows walls to
breathe. 4154-01

5 **TUFTED WOOL CARPET**
An innovative range of
carpeting that utilizes
unexpected materials,
such as linen, steel cable,
paper, goat hair, and felt.
The carpets are available
in widths up to 6.5 ft/2 m
and in area rugs, with this
particular example being
a cut pile of 100 per cent
wool shag. 0045-10

1

2

3

4

5

**1 LAMINATED WOOD VENEER**
Real veneers laminated to a phenolic core for application to dry, interior, light-duty horizontal or vertical surfaces. Available in a broad selection of natural wood-grain sheets. 246-08

**2 WOOD-FABRIC MULTILAYER LAMINATE**
A combination of multilayered wood and fabric. It maintains the formal and visual qualities of wood while providing the softness and give of fabric. Available in beech, walnut, cherry, blue, gray and orange. 1502-01

**3 LAMINATED WOOD VENEER**
Uniformly thin-core veneers of narrow (1/16 in./0.15 cm) strips of solid-grade alder and birch laminated at right angles. This all-hardwood core produces a strong, lightweight panel with minimal voids. 264-03

**4 FLEXIBLE LAMINATED WOOD VENEER**
Thinly sliced sheets impregnated with natural resin that can also be backed with paper, leather and natural fabrics. They are water- and UV-resistant and available in nine different woods. 1502-02

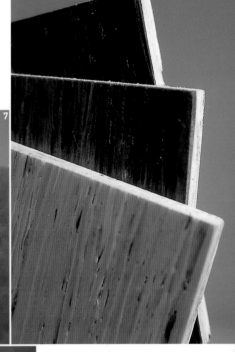

### 5 STRIATED, COLORED WOOD COMPOSITE
Gluing together peeled or sliced natural or dyed woods creates blocks with a marquetry or simulated natural-wood structure. 1856-01

### 6 CERTIFIED WOOD VENEER
Maple, red oak, mahogany and cherry veneers from international certified forestry. Available quartered, rift or plain sliced. 2662-02

### 7 TRANSLUCENT MICA SHEET
Made of mica splittings laminated with shellac resin. The panels are available in amber or silver and in three thicknesses. 3572-02

### 8 STRUCTURAL-COMPOSITE LUMBER
Flooring and decorative surfaces made of structural-composite lumber. This shrink-resistant, formaldehyde-free, engineered wood surface is manufactured of variable-width strands of up to 12 in./30.5 cm in length. 4489-01

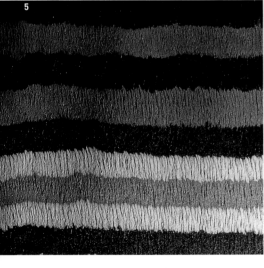

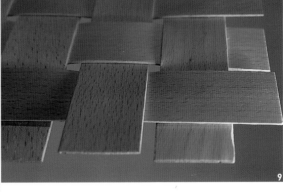

### 9 WOVEN-WOOD SHEETING
Thin wooden slats that are woven together to produce a flexible sheet that can be used for doors, beds and chairs. Available in any type of wood and a wide range of slat widths. 3306-01

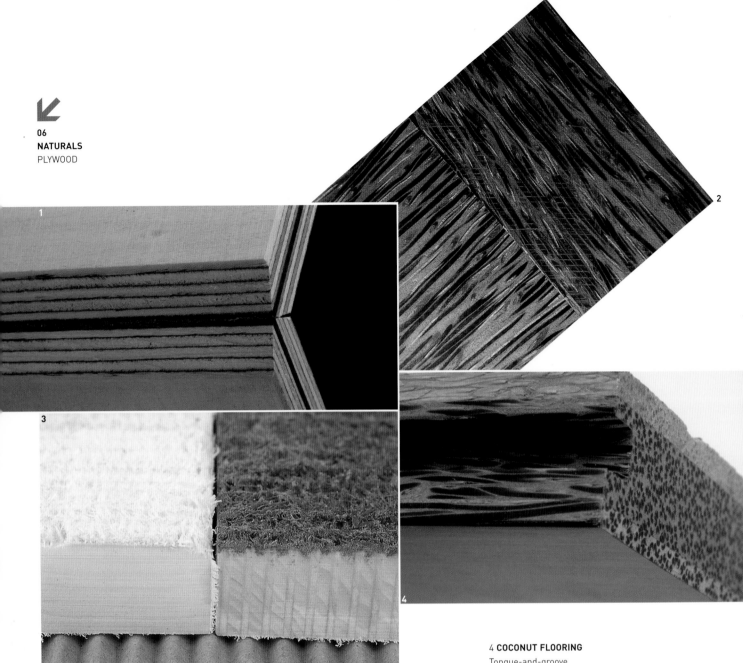

**1 WOOD-LEATHER COMPOSITE**
Allows for easy construction of modular furniture. 5062-01

**2 PALM WOOD-VENEER PANEL**
Brown-on-brown-toned, organically textured, air-dried wood from southern India that is squared up, planed and sanded on both sides. 4645-01

**3 TEXTURED ACOUSTICAL WOOD PANELING**
High-pressure water jets abrade the 10 per cent aspen-wood surface of these panels to produce acoustic dampening. 4670-01

**4 COCONUT FLOORING**
Tongue-and-groove flooring made from 100 per cent coconut palms on plantations that no longer bear fruit. This lumber is on average 60 to 80 years old. 4687-01

5

6

7

#### 5 **VENEER TILE**
To make these tiles, sheets of North American hardwood veneers are pressed together, saturated with epoxy resin and cut to various standard tile dimensions. 4804-01

#### 6 **MOLDABLE PLYWOOD**
Plywood molding for product and furniture design. These curved wood panels are produced from five single beech veneers glued with urea formaldehyde resin using aluminum and steel molding tools at a temperature close to 212°F/100°C. 4824-01

#### 7 **ISOSTATIC COMPRESSED WOOD**
Isostatic compressed wood produces timber with extraordinary hardness, density and durability. Although the cells of the wood are compressed, the wood's structure remains intact. 4738-01

**1 CORK FLOORING**
Floor tiles for commercial and residential use of natural cork, which is the bark of the cork tree. It is durable, resilient, intrinsically impervious to liquids and gases, does not absorb dust, and has thermal and acoustical insulating properties. 1098-01

**2 LAMINATED PEGBOARD**
Laminated pegboard made from melamine, polyvinyl chloride, high-pressure laminate wood veneer, chrome-plated steel, Mylar® or aluminum that is laminated onto medium-density fiberboard (MDF). The material may be custom cut to any specification. 1910-01

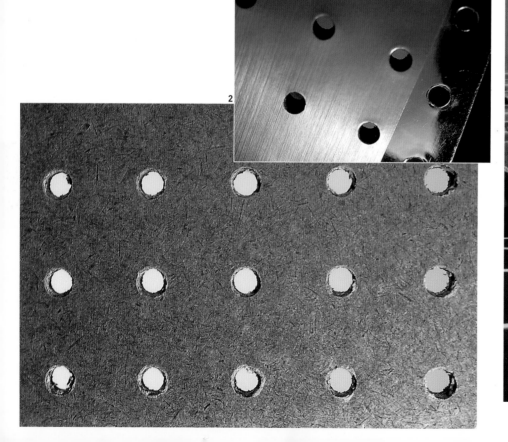

**TOPAKUSTIK**
PALACIO GIUSTIZIA,
VALENCIA, SPAIN
The walls of this
auditorium are clad with
a sound-absorbing panel
system. The slats are
available in three different
widths and spacing
patterns to respond to
wave length. The backs of
the panels are perforated
to allow for the attenuation
of sound waves before they
are absorbed by a soft
backing.

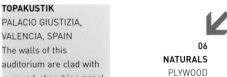

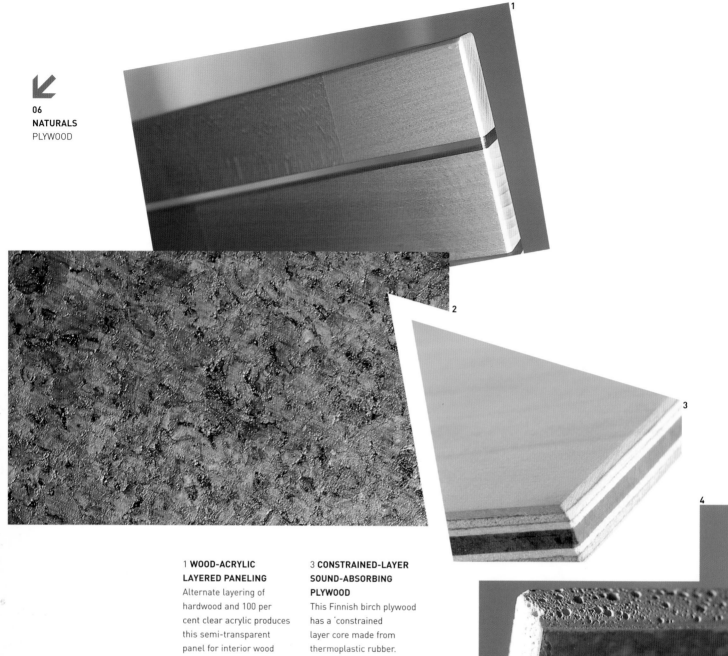

1 **WOOD-ACRYLIC
LAYERED PANELING**
Alternate layering of
hardwood and 100 per
cent clear acrylic produces
this semi-transparent
panel for interior wood
furniture applications.
3305-01

2 **CORK FLOORING**
Natural floor tiles
for commercial and
residential uses. These
tiles are durable, resilient
and intrinsically
impervious to liquids and
gases. They do not absorb
dust and have thermal and
acoustical insulating
properties.  1098-01

3 **CONSTRAINED-LAYER
SOUND-ABSORBING
PLYWOOD**
This Finnish birch plywood
has a 'constrained
layer core made from
thermoplastic rubber.
4868-02

4 **IMAGE-TRANSFER
PROCESS**
A process for transferring
a decorative design. The
resulting patterned sheet
is laminated onto a variety
of flat surfaces and can be
top-coated. Any image can
be transferred in any
combination of colors.
705-01

**5 SALVAGED WOOD VENEER**
Flat-sliced holey veneer from disease-killed butternut wood salvaged from the forest floor.
4683-01

**6 SEA-SHELL MOSAIC TILE**
Mosaic tiles created from discarded shells indigenous to the Philippines that resemble marble in both appearance and durability. Each shell is hand-cut, applied to a ceramic base and polished to a shiny finish. The tiles can be installed on almost any surface. 3602-01

**7 CORK FABRIC**
Fabrics made from natural cork inlaid with dark, toasted cork and laminated to a tough 100 per cent cotton backing. Natural cork is the bark of *Quercus subera*, a Mediterranean oak tree that regenerates every ten years; it is a renewable resource.
1429-03

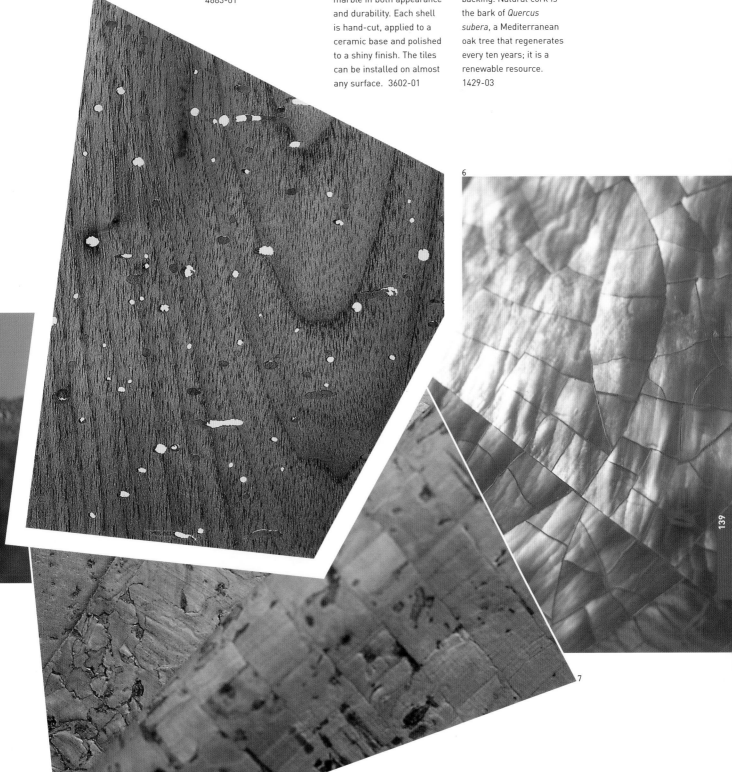

**1 RECYCLED-WOOD MEDIUM-DENSITY FIBERBOARD**
Industrial particleboard with a flat, smooth surface. Made from 100 per cent certified post-manufactured wood, it meets Green Cross Certification standards. 4223-01

**2 RECYCLED ASH-GLASS POROUS CONSTRUCTION FILLER**
A material for soundproofing that is lightweight, strong, fire-retardant and recyclable. It is composed of two residue products: fly ash and granules from recycled glass. 1480-01

**3 RECYCLED PAPER BRICK**
High-strength brick produced from 100 per cent recycled waste 'glossy' paper. A proprietary process that uses high pressure to remove the air and water from the recycled material. 4676-01

**4 RECYCLED-FIBER INSULATION**
Insulation made from recycled, recyclable natural fibers. Composed of post-industrial denim and cotton fibers that are treated with a boron-based flame retardant. This insulation also serves as an anti-fungal agent and is an alternative to fiberglass insulation. 4598-01

**5 RECYCLED CAR-TIRE CARPET UNDERMAT**
Carpet under-cushions made from recycled tires. Composed of 92 per cent recycled tire tread and 8 per cent SBR synthetic rubber with a cellulose fiber/fiberglass backing. 4606-01

**6 WATER-SOLUBLE PACKING PEANUTS**
Reusable, static-free loose fill that cushions and protects just as well as polystyrene. Composed of starch and water and produced by an extrusion process, the peanuts dissolve easily in a small amount of water. 4667-01

**7 FOAMED-POTATO PLATE**
Plates made from recycled potato products. This product is 100 per cent biodegradable, 100 per cent compostable and non-toxic. As rigid as polystyrene and a better insulator, it is suitable for all fast-food applications. 4762-01

### 1 SOY FOAM

Foam from soyoyl, a soy-based polymer compound. Using polyurethane chemistry and standard polymer-production equipment, soyoyl is polymerized and formulated into urethane cellular rigid and flexible foams. 4854-01

### 2 COMPOSTABLE POLYMER

Compostable polymers made from corn and sugar beets. Fermenting the plants' sugars yields the lactic acid from which the polymers are constructed. 3254-01

### 3 INJECTION-MOLDABLE LIGNIN

This natural alternative to thermoplastics may be processed in standard, plastic-injection molding machines. They are composed of 50 to 70 per cent wood chips, 20 per cent ground corn and 10 per cent natural resin and additives. 4612-01

### 4 LIGNIN THERMOPLASTIC

Lignin, found in nature as a structural element of tree trunks, is a by-product of the paper industry. When mixed with natural fibers from flax, hemp or other fibrous plants, it produces a compound that may be processed using conventional molding techniques. 4800-01

**ERIKO HORIKI**
*WASHI* LAMPS
These lamps made with a fibrous paper shell are based on *washi*, a fifteen hundred-year-old traditional Japanese papermaking method. Water is thrown onto the sheets from different angles, creating the organic fibrous patterns.

**5 QUARTZ-AGGREGATE SURFACING**

A dense, strong, durable, non-porous and stain-resistant surfacing material composed primarily of quartz aggregate (93 per cent). Pigment and a proprietary resin binder account for the remaining 7 per cent. 4383-01

**6 VOLCANIC MOSAIC TILE**

Volcanic stone from Southern Europe has been polished flat and glazed to produce this decorative mosaic tiling for indoor applications. 4728-01

**7 RECOMPOSED MARBLE TILE**

Ready-to-lay flooring of finished single slabs and tiles made of recomposed marble consisting of 94 per cent selected crushed marble and quartz granules and 6 per cent polyester resin. 1715-01

**8 LIGHTWEIGHT STONE LAMINATE**

Glueing a thin layer (0.12 to 0.24 in./3 to 6 mm) of natural stone – such as limestone, marble or granite – to a thin layer (0.04 to 1.2 in./1 to 30 mm) of fiberglass-reinforced epoxy creates this lightweight stone laminate. 3638-01

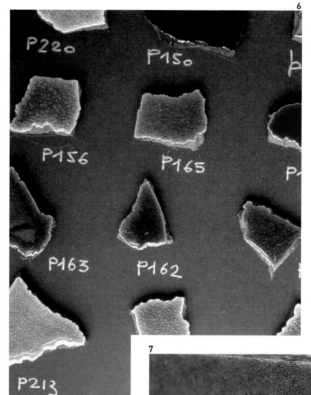

**06**
**NATURALS**
RESIN +
STONE

**1 BIODEGRADABLE POLYMER**

A patented polymer resin made from cornstarch that is processed using heat pressure and water. In an actively managed composting facility, the polymer completely biodegrades into $CO_2$, water and biomass. 4270-01

**2 COMPOSTABLE POLYMER**

Resin recommended for injection-molding, profile-extrusion and sheet-extrusion. It is made of 60 per cent treated cellulose ester, 30 per cent environmentally safe plasticizers, 0.5 per cent colorants and 9.5 per cent fillers. 4688-01

**3 WATER-SOLUBLE POLYMERS**

Biodegradable resin for blown/cast film and injection-molded applications. They are made of 62 per cent PVOH, 3 per cent methyl alcohol, 3 per cent sodium acetate, 2 per cent water, 20 per cent environmentally safe plasticizers and 10 per cent additives. 4688-02

**4 RENEWABLE-SOURCE POLYESTER-RESIN SYSTEM**

This material utilizes 25 per cent (by weight) raw materials derived from soybeans and corn (17 per cent soybean oil, 8 per cent ethanol) combined with 75 per cent petrochemical. 4744-01

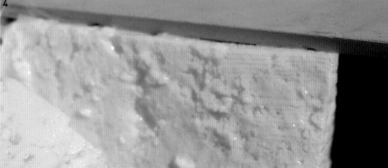

5 **LATENT-HEAT STORAGE WAX**

Paraffin-based storage materials that undergo a phase change at a specified temperature. These phase-change materials are made of highly crystalline paraffin/wax substances bound to a solid granulate. 4756-01

6 **STARCH-BASED POLYMER**

Potato-starch granulate for conventional injection-molding and extrusion machines. 4813-01

7 **CHALK-POLYMER COMPOSITE**

This polymer-film packaging material mimics eggshell chemistry to reduce significantly the amount of material needed to contain a liquid or solid substance. Composed of calcium carbonate (chalk) and polyolefin, it uses 50 to 70 per cent less polymer than standard compositions. 4945-01

1 **DECORATIVE STONE COMPOSITE SURFACING**
A surfacing material made of colored glass plus a synthetic stone. Granules of a proprietary synthetic stone, Avventurina, are mixed with colored glass and an organic resin binder to produce thin slabs of a solid surfacing material. 1679-03

2 **QUARTZ-COMPOSITE SURFACING**
This dense, non-porous and stain-resistant surfacing material consists of 93 per cent quartz aggregate, 7 per cent pigment and a proprietary resin binder. It is strong, durable and tough. 4383-01

3 **MARBLE-CHIP TERRAZZO TILE**
Pre-cast tiles and slabs of 95 per cent marble chips bonded together with about 5 per cent by weight of polyester resin, plus inorganic metal oxides to make them colorfast and resistant to fading or yellowing due to exposure to UV radiation. 2677-01

4 **QUARTZ-AGGREGATE SURFACING**
Quartz aggregate with a durable, colorfast, ceramic color coating. It is unaffected by UV radiation and most chemicals, and is produced in two grades: the S-grade, a spherical aggregate available in twelve colors, and the T-grade, a crushed angular aggregate available in five colors. 4395-01

4

2

3

1

**EFFE MARMI SPA**
RIVERSTONE
A transparent or pigmented polymer matrix within which marble pebbles are suspended. It can be molded into custom shapes and comes in tiles of various dimensions off the shelf. An ideal product for wet areas.

**5 MARBLE-COMPOSITE SURFACING**
Tiles and slabs composed of 95 per cent marble chips and 5 per cent polyester resin. The tiles are available in polished, honed or sandblasted finishes in five sizes, and with a cushion edge.
90-03

**6 MARBLE-AGGREGATE TILE**
Pre-cast cement tiles produced in two layers. Marble aggregates, marble dust and high-compression white cement make up the first layer. Crushed rock, sand and gray cement are vibrated and hydraulically pressed to form the second layer.
1403-01

**7 SLATE-POLYMER COMPOSITE**
An economic replacement for slate roofing materials made of ground slate cast in molds to give the appearance of natural slate. It is lightweight, does not absorb moisture, has a Class A fire rating and comes with a fifty-year guarantee.  3635-01

5

6

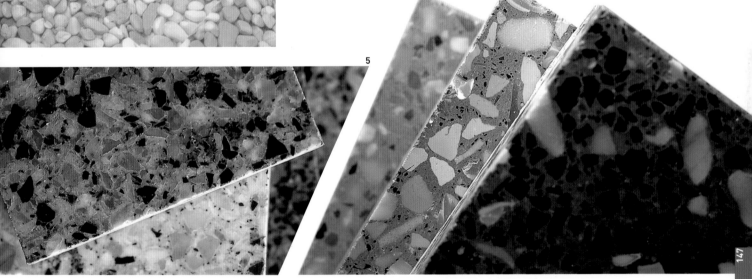

7

2 **COTTON-PAPER FLOORING**
Soft to the touch, these 100 per cent cotton-and-paper-yarn floor coverings are durable, dust-repellent and washable. Available in natural and sixteen colors. The rugs are available in 2.9 by 6.6 ft/88.4 by 201 cm standard and also in custom sizes.  1641-01

4 **WOVEN-PAPER TEXTILE**
Textiles woven of 100 per cent spun paper yarn for upholstery, blinds and partitions, and of 86 per cent paper yarn and 14 per cent cotton for carpets. 1630-01

5 **HEMP CASEMENT**
High-quality hemp fiber is manufactured into yarn to produce this loose-weave casement designed for window dressings. 4193-01

6 **CREATION BAUMAN**
TEXTILE
A translucent, decorative drapery textile, this fabric can be used for shading or screening. The delicacy of the weave makes it more suitable for low-impact uses.

1 **HORSEHAIR UPHOLSTERY FABRIC**
Fabric of horsehair woven with cotton or silk on a specially designed loom. Available in bright colors and black, patterning including stripes, and embellished with custom embroidery.  1594-01

3 **FLAME-RETARDANT UPHOLSTERY FABRIC**
Compostable upholstery fabric composed of 46 to 48 per cent ViscoseFR (a renewable cellulose fiber) and 52 to 54 per cent wool. 3419-02

3

1

2

4

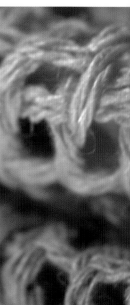

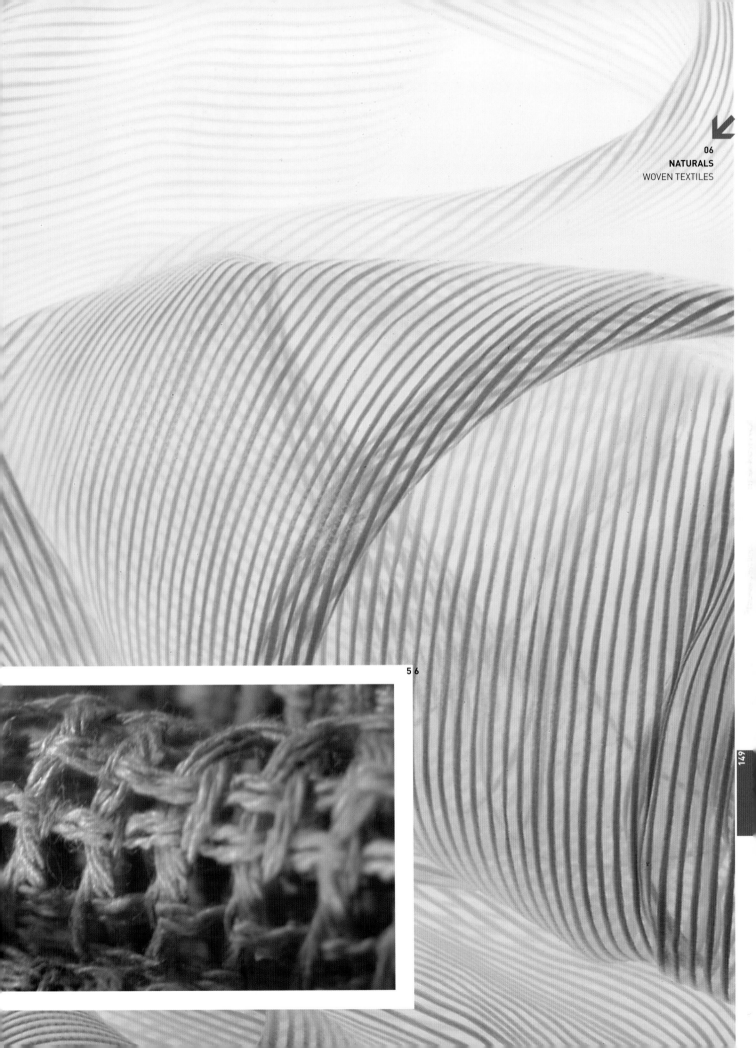

5 6

1

2

3

### 1 TEXTURED FELTED FLOOR TILE
Three-layered tiles comprising an upper layer of mixed nylon (7 per cent) and Teflon®-treated wool, a central section of expanded closed-cell, high-density polyethylene and a lower layer of knit polyamide Velcro®. 4282-01

### 2 KNITTED RESIN-FIBER FABRIC
A fabric specifically designed for tension structures. Millions of tiny holes significantly reduce the temperature underneath the fabric as well as block 91 to 95 per cent of UV radiation. 4409-01

### 3 KRAFT-PAPER FABRIC
Woven fabric created from kraft-paper yarns and tapes. Derived from cellulose fibers and woven Scandinavian kraft paper, these fabrics are colored using water-based dyes and can be spray- or dip-treated to be non-fraying and cleanable. 4873-01

## 4 TACKABLE SELF-HEALING WALLBOARD

A washable, tackable surfacing material with self-healing properties that is durable, has low light-reflectance and doesn't warp or crumble. It is available in thirteen solid colors and also can be used as a decorative finish for furniture and doors. 3895-02

## 5 PAINTED-CORK WALL COVERING

Mixing powdered mica with water-based inks derives the aluminum and copper pigments that create a shimmering, textured and irregular surface over the cork sheets. The pigments come from natural and renewable resources. 4693-01

## 6 MOLD-RESISTANT WALL COVERING

A breathable, decorative wall covering made of polyester, acrylic polymers and cellulose. It minimizes the formation of mold and mildew by allowing moisture to evaporate. It is Class A fire-rated for commercial use. 3896-01

## 7 SISAL WALL COVERING

An almost indestructible, sound-absorbent wallcovering made of 100 per cent sisal. After vat-dyeing and bouclé weaving this natural fiber, it can be installed seam-free over uneven wall surfaces such as concrete block. Offered in twenty colors. 0053-01

## 8 WALL PLASTER

With a suitable primer, this plaster paste can create different effects when applied to most common wall and flooring materials, from marble chip to wood. It is made of a mixture of resins and quartz/silica micro-particles. 1528-03

Polymers can be defined as naturally occurring or synthetic compounds consisting of large molecules made up of a linked series of repeated simple monomers. The materials within this category are predominantly petrochemical-based polymers, meaning that they are synthesized from oil and have carbon atoms as the backbone of their linked molecules. Polymers are also known as resins or plastics. The terms *polymer* and *(thermo) plastic* are often used interchangeably; however, thermoplastics are polymers that can be softened and rehardened when heated and cooled, while thermosets are polymers that require heat, ultraviolet light or a catalyst to cross-link and harden, and will decompose rather than melt when reheated.

For at least the last forty years, of all the areas of new-materials research, the field of polymers has stood head and shoulders above the rest in terms of new products, improved properties and sheer versatility of application. Tougher, lighter, cheaper and more easily processible polymers are solving many materials problems once thought to be the exclusive realm of metals or glass. The sheer number of formulations is staggering, with new versions developed daily.

The widening range of applications is due in part to the numerous forms that polymers can take. From the base resin, there is no limit to the type of form that may be produced, including fibers, monofilaments, yarns, woven and knitted textiles, nonwovens and felts, sheets, films, slabs and bars, as well as any number of injection-molded, extruded, blow-molded, vacuum-formed and rotational-molded parts. The rubber-like feel and elasticity of elastomers, the incredible optical clarity of polycarbonates, the stiffness and toughness of nylons, the weathering- and soiling-resistance of fluoropolymers, the recyclability of polyesters, the

**AQUA GALLERY**
AQUACREATIONS
Aquacreations use four-way-stretch mesh fabric over a steel-wire frame to create unique organic shapes. Inspired by flora and fauna, these lamps are at the scale of the human body.

durability of vinyls and the processability of olefins such as polypropylene and polyethylene – there are families of polymers for every kind of application need. In addition, there are virtually no color limitations, with many also available in transparent and degrees of translucency. The market for 'special-effect' pigment additives to polymers is massive, with notable breakthroughs in metallic, pearlescent (page 187/5), interference, brightening and color-changing (page 187/6) additives. Add to these the dramatic improvements in the longevity and intensity of photoluminescent (glow-in-the-dark) additives, as well as edge-bright and fluorescent pigments, and the range becomes almost limitless.

Among the many new formulations and applications of polymers that continue to emerge are those which herald a new way of thinking that makes us look again at their supposed limitations and marvel at their evolution. The more recent and promising results of this evolution have included the use of nano-composite additives to resins and coatings, the visual and tactile impact of gels (page 204/1), and the metal-like behavior of conductive polymers.

Nano-composite polymers' remarkable properties rely on the effect that ultrafine dispersions of platelets have when dispersed within the plastic. These 50 Å/5 nm-thick alumino-silicate clay plates, often hundreds of nanometers long, offer strengthening and surface-

gloss improvements far above what would
be expected for their volume (nominally
2 to 5 per cent of the total part). Initial use
of these polymers was based upon their
improved physical properties such as
impact strength, modulus and
dimensional stability. However, it was
found that these additions increase heat-
distortion temperatures and flame
retardance while improving gas-barrier
properties and optical clarity. The
improvement in clarity has led to the use
of such clay platelets in scratch-resistant
coatings for visors and lenses.

Despite the considerable drawbacks of
excessive weight and limited
processability, polyurethane gels continue
to entice designers due to their uniquely
human-like feel and distinctly alien
appearance. Developed originally for use
in orthopedic mats to maintain blood flow
during lengthy surgical procedures, the
shear behavior of these liquid/solid
materials has lent itself to a host of other
cushioning applications. Further
improvements in surface texture, the

weight issue and processing methods are
likely to bring additional areas of use for
such appealing materials.

Electrical conductivity has long been a
property exclusive to metals. However,
with the discovery in the 1980s of certain
polymers based on polyaniline that were
able to act like metals in their conductivity
of electricity, a host of new applications
arose that utilize this property. Conductive
fibers and textiles; flexible, moldable
plastic batteries that have improved
energy-to-volume (power-to-size)
compared to standard lithium batteries;
printable integrated circuits;
electroluminescent devices and organic
LED's are some examples.

Given this cornucopia of material forms
and types, it is not surprising that this
category contains by far the most
materials within the Material ConneXion
Library, and that this is the largest section
in this book. Be aware that this is merely
the tip of the iceberg!

2

3

07
POLYMERS

4

**07
POLYMERS**
ACOUSTIC + VIBRATION MATERIALS +
COATED TEXTILES

**1**

**1 NONWOVEN SOUND-ABSORPTION PANEL**
These panels have good heat-resistance properties. They are composed of two layers, a p-aramid (Kevlar®) paper and a nonwoven batting made from 30 per cent aramid fibers and 70 per cent PET (polyester) staple fibers. 5011-01

**2 PRINTED VINYL UPHOLSTERY**
Abrasion- and scuff-resistant, this material is composed of 100 per cent virgin vinyl with a 100 per cent polyester loop-knit backing and an anti-microbial, antibacterial finish. 346-01

**3 STAIN-RESISTANT TENT TEXTILE**
Woven glass fabric coated with polytetrafluoro-ethylene. Non-combustible, with good weather, temperature, aging and chemical resistance, this fabric also has a low coefficient of adhesion, thus making it resistant to soiling. For construction engineering. 2288-02

**4 HIGH-STRENGTH TENT TEXTILE**
Dimensionally stable, weather-resistant and durable synthetic fabrics designed and manufactured using PRECONTRAINT FERRARI® technology, which provides a continuous and controlled tension in the weft throughout the coating, calendering and heat-setting process. 1440-01

**5 WEATHER-RESISTANT TEXTILE**
A low-wick polyester textile coated with polyvinylidene fluoride that is weather- and cold-resistant (-40°F/-55°C) and has high breaking- and tear-strength. Available in maroon, green, white and custom colors. 2782-01

**3**

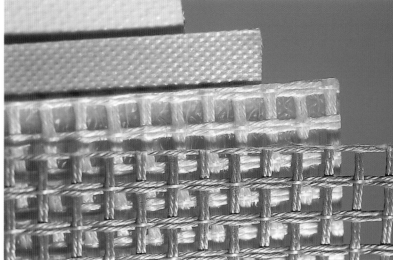

**2**

**4**

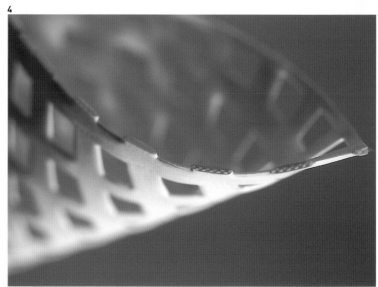

## 7 RETROREFLECTIVE TEXTILE

A material composed of wide-angle, exposed retroreflective lenses bonded to a durable 65 per cent polyester/35 per cent cotton backing, or a flame-resistant 94 per cent polyester/6 per cent nylon tricot knit, or 100 per cent cotton.  2615-01

## 8 DURABLE COMPOSITE FABRICS

Nylon and polyester fabrics, custom-coated with thermoplastics, such as polyvinyl chloride, Hypalon® and polyurethane. For use in the transportation, automotive, aerospace, architectural, energy, military and marine industries.  2786-01

## 6 HIGH-FRICTION SURFACE

This molded, non-PVC elastomeric thermoplastic consists of thousands of tiny 'fingers' that provide a strong, secure grip under both wet and dry conditions.  2615-08

6

7

5

8

**07
POLYMERS**
COATED TEXTILES

**1 TENT MEMBRANE**
A family of translucent architectural membranes that are durable, stain-resistant and energy efficient. Composed of fiberglass coated with polytetrafluoroethylene. For use in permanent structures. 2915-01

**2 RETROREFLECTIVE SPORTSWEAR TEXTILE**
The matrix of millions of reflective 'dishes' is incorporated into the textile to reflect light back to the original source. 3502-01

**3 ABRASION-RESISTANT COATED FLEECE FABRIC**
Non-slip and abrasion-resistant 100 per cent PVC coating laminated onto 100 per cent polyester fleece fabric. 3749-03

**4 DYE-SUBLIMATION DIGITAL-PRINTING PROCESS**
This color-printing process entails reproducing a digital image onto a display surface. Dye sublimation is a four-color process transferrable onto all non-natural fabrics, wood, metal and plastics. 4136-01

**5 SILICONE-COATED FIBERGLASS BELTING**
Belts made of a woven fiberglass cloth substrate coated on one or both sides with silicone. They have good release properties, high dielectric strength and high wear, flex and chemical resistance. In white or red. 1485-04

**6 SHAPE-MEMORY POLYMER TEXTILE**
A laminate for textile applications. Above a specified activation temperature, this polyurethane-based membrane acts as a waterproof layer that allows heat and condensation through. 4562-01

3

1

2

**7 COATED UPHOLSTERY TEXTILE**

Synthetic lizard-skin textile composed of PVC with a cotton-polyester blend backing. 4595-01

**8 COATED UPHOLSTERY TEXTILE**

A custom-manufactured textile composed of 95 per cent PVC on a 5 per cent polyurethane substrate in ivory, tan and black. Available in 55 in./140 cm widths. 4595-02

5

6

4

7

8

**1 SILICONE-COATED TEXTILE**
These silicone-treated materials can be designed using almost any color, scale, texture and fabric, and are custom-designed in small runs of up to 20 yds/18 m, with a minimum of 1 yd/0.9 m. 4608-01

**2 ANTI-BACTERIAL TEXTILE**
A textile containing catechin, a substance extracted from the shells of crabs and shrimp that has inherent odor-preventing and antibacterial properties. 4741-02

**3 WEAR-RESISTANT METALLIC TEXTILE**
This high-tech textile composed of 77 per cent polyester and 23 per cent polyurethane withstands 102,000 double rubs under the Wyzenbeek test. It is water-resistant, as well as bleach- and stain-resistant. 4760-01

**4 SILICONE-COATED FABRIC**
Custom-designed in small runs of up to 20 yds/ 18.3 m, with a minimum of 1 yd/0.9 m. These very creative silicone-treated fabrics can be designed using almost any color, scale and texture. 4608-01

**5 POST-INDUSTRIAL RECYCLED PANEL FABRIC**
An environmentally friendly panel system and upholstery fabric manufactured with Terratex®, a 100 per cent postindustrial recycled polyester. It withstands 69,000 double rubs and is available in eleven different colorways composed of eight different colors. 4760-03

1

2

3

4

5

7

6 **CUT-RESISTANT TEXTILE**
Abrasion- and cut-resistant layered fabric comprising an inner layer composed of an elastomeric fabric and an outer network of hexagonal platelets that are able to move with respect to each other, thus offering a continual protective surface. 4967-01

7 **MOLDABLE RUBBER MEMBRANE**
These durable membranes are made of ethylene propylene diene monomer rubber. They are flexible at both high and low temperatures, resistant to tearing, chemicals and microorganisms, and can be homogeneously joined together without the use of solvents. 1616-01

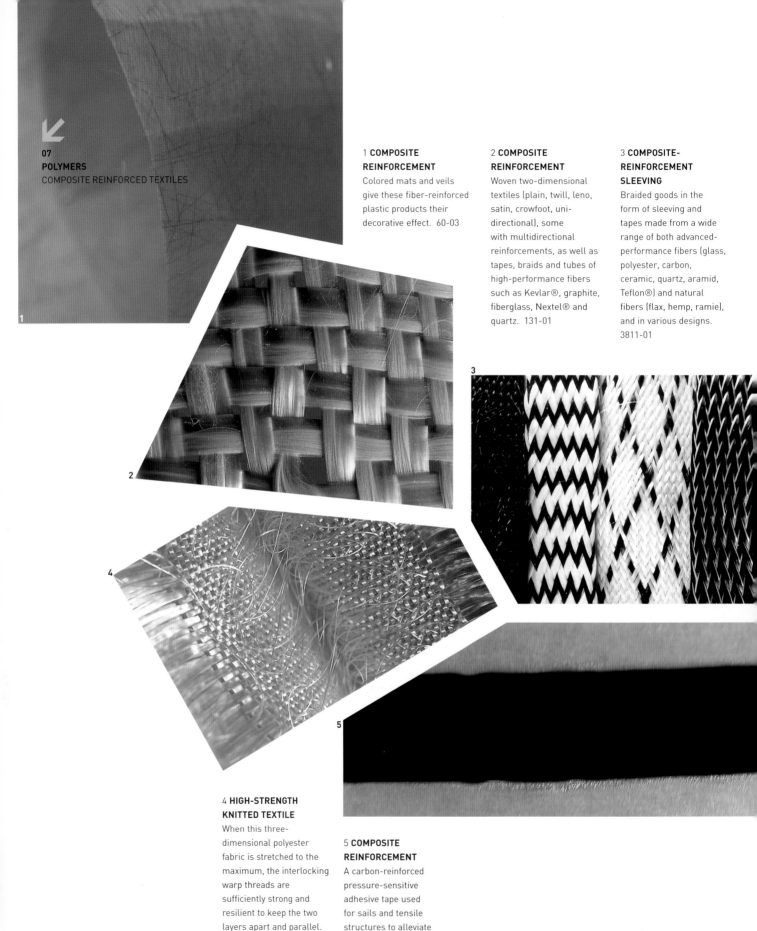

**07
POLYMERS**
COMPOSITE REINFORCED TEXTILES

**1 COMPOSITE REINFORCEMENT**
Colored mats and veils give these fiber-reinforced plastic products their decorative effect. 60-03

**2 COMPOSITE REINFORCEMENT**
Woven two-dimensional textiles (plain, twill, leno, satin, crowfoot, uni-directional), some with multidirectional reinforcements, as well as tapes, braids and tubes of high-performance fibers such as Kevlar®, graphite, fiberglass, Nextel® and quartz. 131-01

**3 COMPOSITE-REINFORCEMENT SLEEVING**
Braided goods in the form of sleeving and tapes made from a wide range of both advanced-performance fibers (glass, polyester, carbon, ceramic, quartz, aramid, Teflon®) and natural fibers (flax, hemp, ramie), and in various designs. 3811-01

**4 HIGH-STRENGTH KNITTED TEXTILE**
When this three-dimensional polyester fabric is stretched to the maximum, the interlocking warp threads are sufficiently strong and resilient to keep the two layers apart and parallel. 2394-01

**5 COMPOSITE REINFORCEMENT**
A carbon-reinforced pressure-sensitive adhesive tape used for sails and tensile structures to alleviate some of the stress in excessive load-bearing areas. 4250-02

6 **ANTI-STATIC FILAMENT**
A fine-filament, bi-component yarn, each filament of which has a trilobed conductive carbon core surrounded by a sheath of polyester or nylon. 3099-01

7 **ACRYLIC LIGHT GUIDE**
Synthetic light guide for display applications. The 100 per cent acrylic rod is co-extruded as a three-part structure consisting of an outer 'light-channeling' coating, an acrylic core and a reflector strip. 5070-01

8 **ELECTROLUMINESCENT WIRE**
A durable, flexible plastic wire that emits light throughout its entire length and operates on either battery controllers or AC, requiring a relatively small voltage. 4453-01

9 **FLAME-RETARDANT ELASTOMER**
A shielding material made of a silicone elastomer filled with nickel-coated graphite particles that protects against electromagnetic interference (EMI) at the same time as it provides high electrical conductivity and broadband shielding. 893-01

7

6

8

9

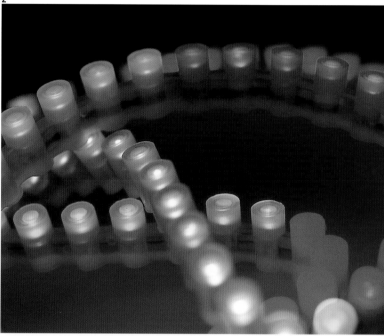

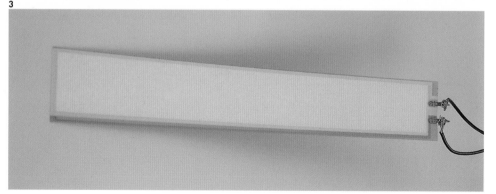

1 **ACRYLIC ADHESIVE SHEET**
Sheet composed of a proprietary adhesive coated on a release paper. It is used to bond flexible details together on multi-layer boards and to bond flexible circuits to rigid board.  3363-01

2 **SILICONE-COATED LED**
Individual white light-emitting diodes have been connected in parallel and immersed in clear, flexible silicone to produce decorative light strips.
4680-01

3 **ELECTROLUMINESCENT POLYMER SHEET**
These flexible sheets are composed of phosphor encapsulated in aluminum oxide and a proprietary polymer chemistry deposited onto an indium/tin-oxide-sputtered polyester substrate.
4656-01

4 **FLEXIBLE PHOTOVOLTAIC PANELS**
Lightweight adhesive-backed shingles that produce voltage when exposed to light. Installed directly over roof sheathing or purlins, these shingles emulate conventional roofing materials in construction, function and installation.  4682-01

**FUSEPROJECT**
SWAROVSKI NY
An electroluminescent
polymer sheet provides an
even light within a draped
Swarovski crystal shade.

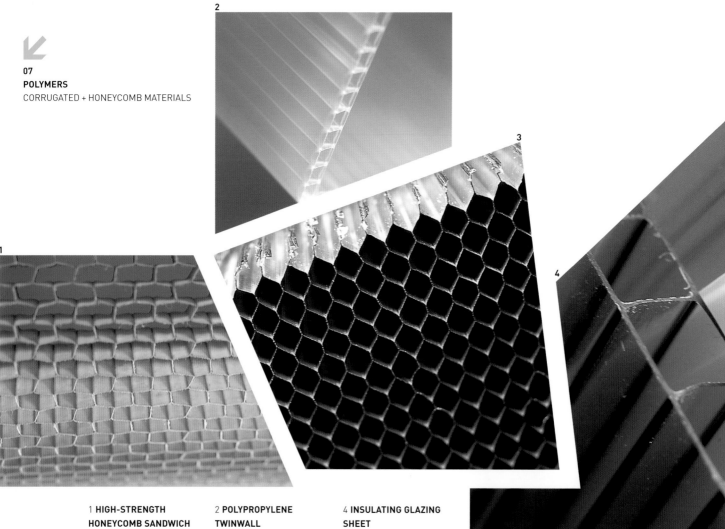

### 1 HIGH-STRENGTH HONEYCOMB SANDWICH PANEL

Panels fabricated by laminating various face materials to honeycomb cores made of aramid fiber, phenolic fiber or aluminum. They meet the stringent fire, smoke and toxicity requirements for aircraft and train interiors. 3462-01

### 2 POLYPROPYLENE TWINWALL

An extruded, corrugated polypropylene sheet produced in two basic forms – fine-flute spacing (0.13 in./3.2 mm) and wide-flute spacing (0.2 in./5 mm) – in a range of weights and dimensions, and in translucent plus a variety of colors.  3509-01

### 3 ARAMID-FIBER HONEYCOMB SANDWICH CORE

A variety of cores made with a range of metal and plastic materials including aluminum alloys, aramid fiber, polypropylene and polycarbonate. The aluminum honeycomb is lightweight but strong and can be used at relatively high temperatures. 119-01

### 4 INSULATING GLAZING SHEET

A lightweight, flexible, weather-resistant, flame-retardant sheet made of extruded polycarbonate which can be worked like wood and has thermal conductivity comparable to insulating glass. Available in bronze, ice and clear. 133-01

### 5 HIGH-STRENGTH HONEYCOMB SANDWICH PANEL

A structural panel that is lightweight and impact- and corrosion-resistant. This board has a high strength-to-weight ratio and has low fire, smoke and toxic gas emission. 1281-01

### 6 LIGHTWEIGHT HONEYCOMB SANDWICH PANEL

Sheets made of various thermoplastics that are lightweight with a high strength-to-weight ratio. 1534-01

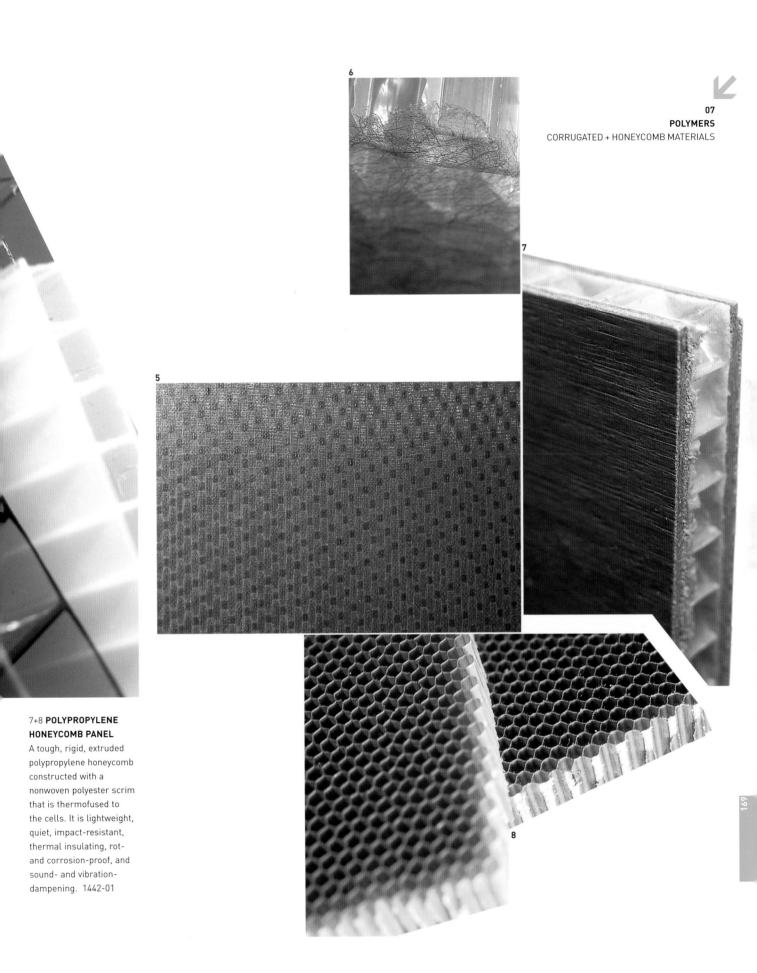

7+8 **POLYPROPYLENE HONEYCOMB PANEL**
A tough, rigid, extruded polypropylene honeycomb constructed with a nonwoven polyester scrim that is thermofused to the cells. It is lightweight, quiet, impact-resistant, thermal insulating, rot- and corrosion-proof, and sound- and vibration-dampening. 1442-01

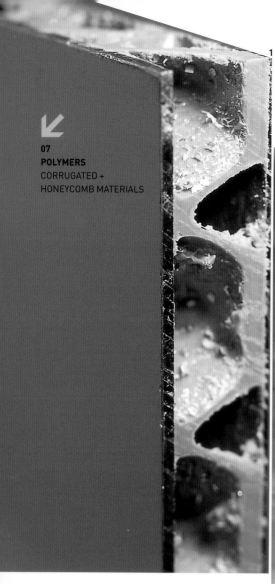

1

2

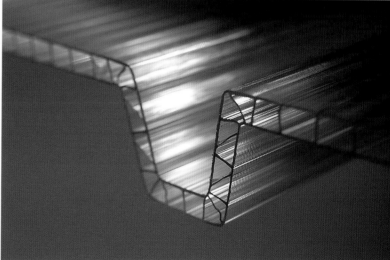

3

1 **ARAMID-FIBER
SANDWICH CORE**
A tough, environmentally
resistant core material
for sandwich panels
made of an aramid fiber
(Nomex®)/phenolic resin
honeycomb. It is self-
extinguishing and water-
and fungus-resistant, and
is a good thermal and
electrical insulator.
1281-02

2 **POLYCARBONATE
SKYLIGHT**
A domed, translucent
skylight of extruded
polycarbonate with a
lightweight cellular
structure that resists
UV radiation, impact and
thermal shock. Fabricated
in lengths up to 40 ft/12 m.
2680-02

3 **POLYCARBONATE
SKYLIGHT**
A UV radiation-resistant,
translucent, ribbed
polycarbonate sheet with
a cellular structure for
skylights. It is fabricated
in lengths up to 40 ft/12 m
and is available in neutral,
opal, smoky and green.
2680-04

## 4 TRANSPARENT HONEYCOMB INSULATION

Designed as part of the glazing system Solera™, this insulation material is composed of a flexible acrylic that exhibits excellent stability in a filtered-UV environment. 4749-01

## 5 FOAM-FILLED HONEYCOMB SANDWICH PANEL

High-strength, lightweight honeycomb for composite panels. 100 per cent polyurethane foam is used to fill a variety of non-metallic honeycomb structures. 4948-01

## 6 HONEYCOMB CUSHIONING

Thermoplastic polyurethane (TPU) elongated hexagons with alternating thick- and thin-walled cells create a honeycomb geometry that gives effective, flexible cushioning for shock absorption and weight distribution. 4795-01

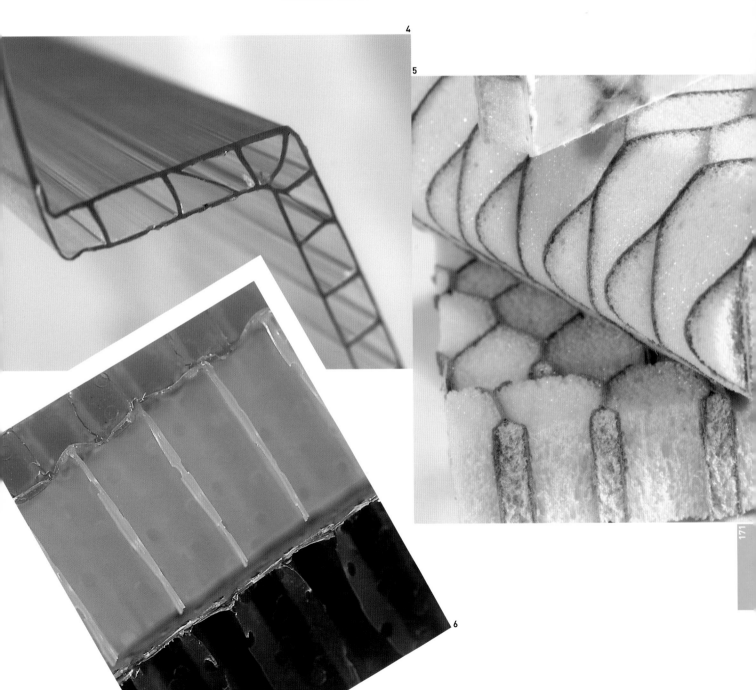

4

5

6

**TIZIP**
ZIPPER
This watertight enclosing device can withstand high-pressure air, forming an extremely effective barrier against the elements and any particulate matter.

**1 FASTENING INTERLAYER**
A clear plastic-film fastener that connects two fabric layers. This thin Velcro®-type material is composed of polypropylene and polyester (or polyamide for higher-temperature uses such as in aircraft) and is fabricated with various types of substrates.
4328-01

**2 HEAT-FUSIBLE YARN**
Hot melt-coated polyester, nylon or fiberglass yarns that are heat-fusible. Fibers are coated with adhesive ethylene vinyl acetate (EVA), polyamide or polyester. Single adhesive monofilaments are also available.
4613-01

**3 SPIRAL WRAP**
This 100 per cent polyethylene spiral sheathing for cable may be expanded and used like tape. It is available in fluorescent colors and black. The black wrap is UV-resistant, allowing it to withstand direct sunlight for extended periods of time. 4622-01

1

2

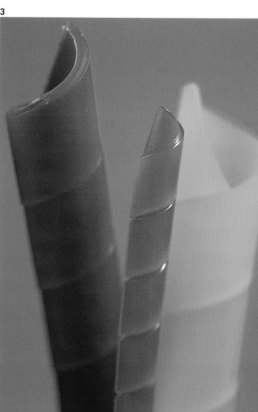

3

**4 MULTIFILAMENT POLYMER TAPE**

Textiles and tapes woven from a variety of different fibers. Multifilaments and monofilaments of polyamide, polyethylene, polyester, aramid and glass are used, as well as copper or stainless-steel wire and cotton and rayon yarns. 4763-01

**5 HIGHLY ELASTIC ADHESIVE FILM**

A polyurethane film with efficient stretch and recovery. The film effectively adheres to polyester, poly/cotton and blended fabrics, thereby providing an alternative to stitching for apparel. 4805-01

**6 AIR- AND WATER-TIGHT ZIPPER**

A strong zipper for high-pressure applications. The zipper teeth are embedded and secured in a plastic profile that seals when closed. Available in four different grades. 4623-01

**7 PUCKER-FREE ADHESIVE TAPE**

Tape used to achieve pucker-free seams. 100 per cent polyethylene with a high-viscosity polyolefin adhesive is adhered to the inside of a seam to achieve pucker-free joins. 4805-02

5

7

4

6

173

1 **BALLISTIC COMPOSITE SHEET**
A strong fiberglass laminate that is bullet-resistant. This woven panel is lighter than steel, ricochet-resistant, has a building material Class 1-A fire/smoke rating, and can be readily machined with ordinary hand tools. 426-01

2 **GLASS-FIBER LAMINATE SHEET**
A translucent laminate (flat or corrugated panels) that is lightweight, strong and shatter-resistant. It is composed of an acrylic-modified polyester resin (containing the panel colors) reinforced with high-strength glass fibers. 799-02

3 **LIGHTWEIGHT STRUCTURAL PANEL**
A strong material that is corrosion-free. It is made of a thermosetting-resin closed-cell foam reinforced with long glass fibers that are homogeneously dispersed. This synthetic wood has a flexural strength comparable to natural wood. 4117-01

4 **GLASS-FIBER COMPOSITE PANEL**
Fiberglass-reinforced multipurpose sheets for architectural applications that are available in a large number of variations. 113-01

5 **GLASS FIBER-REINFORCED POLYMER GRATING**
Corrosion-resistant, non-conductive, non-sparking molded-plastic grating in a square mesh pattern, designed as stair treads. It is only half the weight of welded steel grating of equivalent strength. 168-01

6 **PULTRUDED COMPOSITE PROFILE**
Fiberglass-reinforced polyester pultrusions available in four grades of heat-resistance, including one intended for transformer and similar electrical applications and one that retains its physical and electrical properties at temperatures up to 428°F/220°C. 160-02

7 **TEXTURED VERTCAL SURFACING PANEL**
This sheet material is composed of a proprietary blend of mineral particulates and the polymer compound poly(methyl methacrylate). The sheets have high scratch, abrasion and heat resistance. 243-02

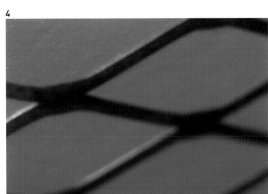

4

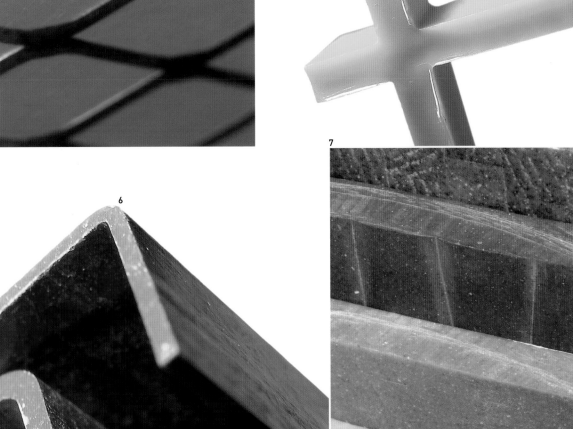

5

6

7

### 6 NATURAL FIBER-REINFORCED POLYMER

Thermoplastic composites reinforced with natural fibers such as coir, jute, kenaf, flax, agave, sisal, bagasse, wood, rice hulls and sunhemp. 3000-01

### 7 HIGH-PRESSURE LAMINATE

Manufactured from paper reinforced with phenol, this rigid, moisture-resistant laminate works well as an electrical insulator in places where humidity conditions are not severe. 3133-01

### 1 EXTRUDABLE WOOD/POLYMER COMPOSITE

A composite plastic material that uses compounded, pelletized wood or paper fiber as a feedstock for injection, extrusion and thermoform molders. 1998-01

### 2 POLYESTER SOLID SURFACING

A polyester resin bonds a mixture of salvaged aluminum chips or other non-ferrous metal waste and a silica filler and pigments to produce this material suitable for countertops, furniture and architectural surfaces. 2697-01

### 3 FIBERGLASS-REINFORCED PULTRUDED GRATING

Lightweight, fiberglass-reinforced, pultruded plastic gratings intended for walkways, platforms and stairs. They are low-maintenance, corrosion-resistant, non-conductive, structurally stable and environmentally safe. 3027-04

### 4 ELECTRICAL-GRADE HIGH-PRESSURE COMPOSITE

A rigid, moisture-resistant, tough composite having good machining characteristics. Applications include switchboard-panel circuit-breakers, switch arms, terminal blocks and motor bases. 3133-02

### 5 THERMOPLASTIC ROVING

Roving for making continuous-fiber composites. The filaments are composed of fine, homogeneous, commingled, unidirectional polypropylene and glass fibers that are twisted in to the roving. 3837-01

**TRESPA**

VIRTUON TEXTUREN
These thick-core laminate sheets are virtually indestructible. The surface of phenolic resin-impregnated kraft paper can be colored, patterned and textured. The material's through-thickness is so hard that the sheets can be cut with great precision, allowing them to be butt-jointed with an almost seamless appearance.

1

**1 POLYMER-MATRIX COMPOSITE**
Fiber-reinforced composites (an epoxy, polyester, vinylester or antibacterial resin reinforced with fibers of glass, carbon and aramid) make up these lightweight profiles and tubes (0.12 to 12 in./3 to 300 mm in diameter). 3877-01

**ECCO**
PERSONAL POND
The Personal Pond is composed of fiber-reinforced plastic, natural stone and laminated textiles. Meant to channel the sensory experience of driving a luxury car, the console references the materials palette found in an automative interior. The tactile feedback from the materials and gesture-sensors translates body language into commands that control an audio sequence and lighting pattern.

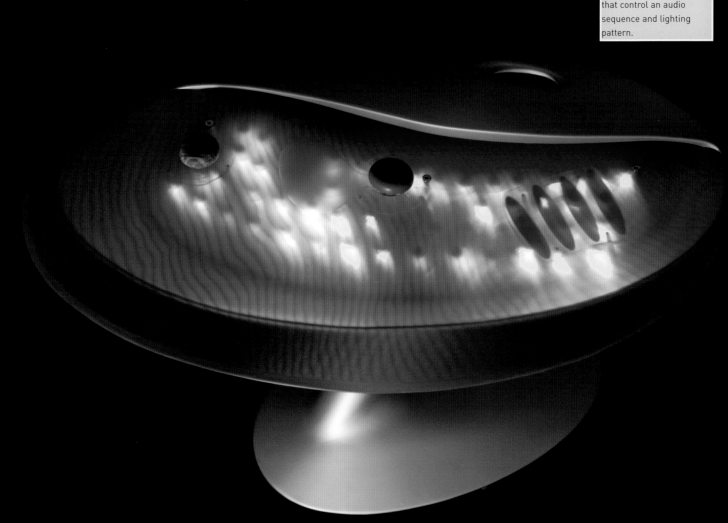

2 **BRAIDED ROPE**
A double-braided rope that is more flexible than most high-modulus fibers and has a high dielectric strength and melting point (621ºF/327ºC). It consists of a polyester sleeve covering an inner load-bearing core of Vectran®, a liquid-crystal polymer. 0168-01

3 **SEATBELT WEBBING**
This webbing, which stretches by selective fiber relaxation, consists of a co-polymer fiber made from polyester-caprolactone block. Intended for use in conjunction with an airbag for passenger cars, the webbing holds the passenger in position at impact, then slowly lets go. 3477-04

4 **PHOTOCHROMATIC THREAD**
Embroidery thread that changes color in sunlight. Specifically, when exposed to sunlight or some other source of UV radiation, white thread changes to one of seven colors: yellow, red, orange, magenta, teal blue, purple or green. 4326-02

5 **WATER-SOLUBLE YARN**
A yarn that is 100 per cent dissolvable in water. This polyvinyl alcohol (PVA) fiber dissolves completely when immersed for about thirty minutes in acidified (pH 5–6) water at about 194º F/90ºC, followed by rinsing in water at about 104ºF/40ºC with agitation. 4469-01

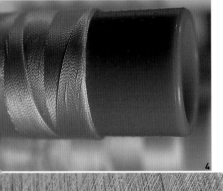

**INGO MAURER**
TROZDEM
This table, made of Corian®, is lit from the center, below eye-level, and has adjustable reflectors to cast rays of light on the seamless cast surface.

**BOFFI (MARCEL WANDERS)**
GOBI
Consisting of a bathtub and two basins, Gobi is made of cast polyester resin.

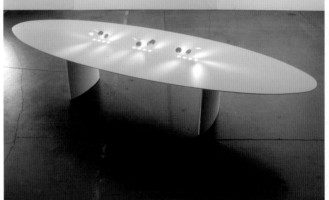

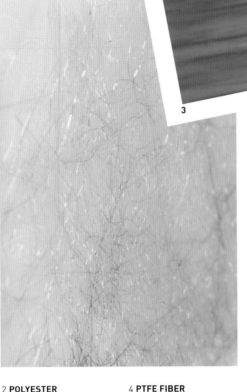

**1 POLYPROPYLENE
FIBRILLATED TAPE**
A translucent tape yarn
made of polypropylene.
The properties and
performance of this
lightweight yarn are
between those of a ribbon-
type yarn and a filament
yarn, with medium to high
shrinkage and moderate
elongation. 150-01

**2 POLYESTER
IRIDESCENT FIBER**
Soft, fine staple fibers
that, when heated to
approximately
225°F/107°C, bond to
each other to produce
nonwoven fabric. The
fibers may also be blended
with other natural and
synthetic fibers. 1466-04

**3 HIGH-STRENGTH
ARAMID FIBER**
Combining high strength,
toughness and thermal
stability, this fiber has
43 per cent lower density
than fiberglass and is five
times stronger and much
more flexible than steel of
the same weight. 2164-02

**4 PTFE FIBER**
Fluorocarbon
(polytetrafluoroethylene)
fibers have a low friction
coefficient and are
chemical-, sunlight- and
weather-resistant. Good
insulators, they remain
stable over a wide range of
temperatures. 2164-01

**5 LIQUID-CRYSTAL
POLYMER FIBER**
A high-performance fiber
with good mechanical
properties which are
maintained over a wide
range of temperatures.
It also has good chemical
resistance and low
moisture absorption.
Applications include ropes,
cables and cut-resistant
clothing. 149-01

**6 HIGH-STRENGTH
ARAMID FIBER**
An aramid fiber that
combines high heat
and flame resistance with
good textile properties.
Applications include
fire-retardant clothing,
electrical insulation,
armor and reinforcement
for rubberized belts and
hoses. 2164-03

8 9

6

#### 7 ORIENTED POLYETHYLENE FIBER

Extremely strong fibers that are eight to ten times stronger than steel of comparable weight.These extended-chain, ultra-high-molecular-weight polyethylene fibers are also light enough to float, highly abrasion-resistant and resistant to chemicals. 3477-01

#### 8 REGENERATED CELLULOSE FIBER

Soft viscose rayon fibers made from 100 per cent regenerated cellulose, a halogen-free flame retardant and phosphor-sulphur-containing pigment. The fibers remain stable at 239ºF/115ºC for up to one hour.  2499-01

7

#### 9 WOVEN OPTICAL-FIBER TEXTILE

Optical fibers composed of a polymer have been woven using a proprietary process into textiles such as Lycra®, nylon and cotton for decorative effect. The density, color and pattern of the fibers may be varied.  4836-01

#### 10 ARTIFICIAL GRASS

A UV-resistant artificial grass surface. The soft 2.5 in./6.4 cm-deep polyethylene and polypropylene blades are tufted to look like grass and 'planted' in a resilient 'soil' made from silica and a material produced from recycled tennis shoes. 4634-01

10

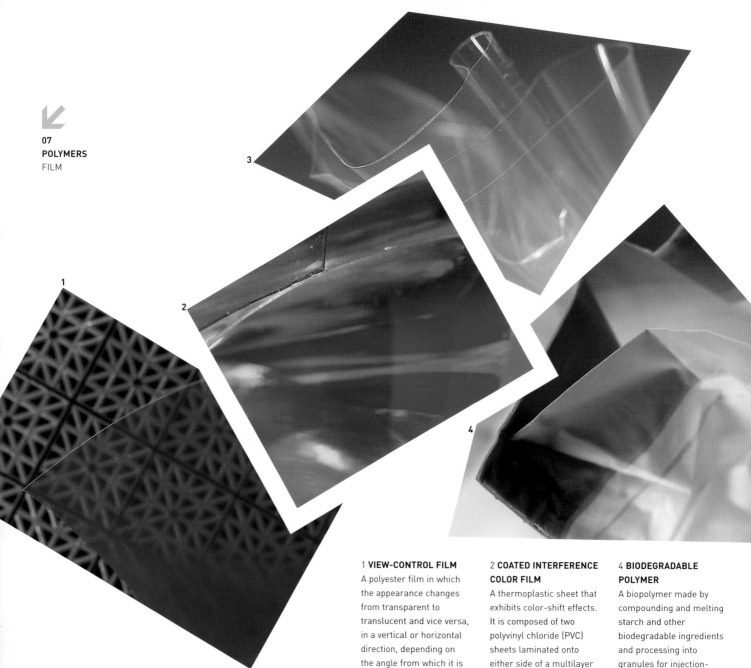

**1 VIEW-CONTROL FILM**
A polyester film in which
the appearance changes
from transparent to
translucent and vice versa,
in a vertical or horizontal
direction, depending on
the angle from which it is
viewed.  1340-04

**2 COATED INTERFERENCE
COLOR FILM**
A thermoplastic sheet that
exhibits color-shift effects.
It is composed of two
polyvinyl chloride (PVC)
sheets laminated onto
either side of a multilayer
film consisting of more
than three hundred
polyester/acrylic layers.
The film's construction
causes interference effects
in the sheet.  2604-08

**3 FLEXIBLE PACKAGING
FILM**
Films for flexible barrier
packaging composed of
ethylene vinyl alcohol
co-polymer (EVOH) and
available in several grades
for both food and non-food
packaging. They provide a
high barrier to oxygen and
other gases, as well as to
aromas and flavors.
3561-01

**4 BIODEGRADABLE
POLYMER**
A biopolymer made by
compounding and melting
starch and other
biodegradable ingredients
and processing into
granules for injection-
molding and extrusion into
flat films and sheets.
3682-01

**5 DECORATIVE COATED
FILM**
Films coated with colored
beads and flakes. Various
colored polyvinyl chloride
beads and/or flakes are
applied to the surface of a
PVC film 17 in./42 cm wide
and 0.02 to 0.03 in./0.05 to
0.07 cm) thick.  3901-03

5

**HONEYCOMB
CUSHIONING**
SUPRACOR HONEYCOMB
The hexagonal design
incorporates sides of
different thicknesses and
rigidities. Each hexagon
collapses in a controlled
manner when pressure
is applied, resulting in
superior cushioning
properties. This material
is ideal for both seating
and shoe soles.

1

**4 NON-METALLIC REFLECTIVE FILM**
Flexible color films made by combining precise multilayers of polymer materials that have different reflective properties with outer layers of polyester. 445-701

**5 REFLECTIVE FILM**
These permanent, self-adhering films for use in exterior applications are available in seven colors and in different designs and levels of reflectivity for architectural, transportation and general signage applications. 4402-01

**3 IN-MOLD DECORATION PROCESS**
A process that permanently incorporates three-dimensional graphics with a plastic part during the injection-molding process. A decorated polycarbonate film is formed into a three-dimensional shape and die-cut. 4239-01

**6 NON-METALLIC MIRROR FILM**
A flexible film that reflects light like a mirror. It has a very shiny, silvery appearance that specularly reflects more than 98 per cent of the visible light that hits its surface from any angle. 4457-02

**1 LENTICULAR FILM**
A display that presents animation with continuous movement. This process shows about three seconds of a film or video. Scenes can be either a continuous-image playback or a series of shorter sequences edited together as a composite. 4057-01

**2 PHOTOLUMINESCENT FILM**
This safety signage comprises safety-grade strontium oxide aluminate pigments that glow in total darkness for at least eight hours and recharge in fifteen to twenty minutes once the light is restored. 4202-01

3

4

5

6

3

1

2

1 **VAPOR-RETARDATION FILM**
A membrane for housing insulation, this nylon 6/caprolactum-blend polymer film replaces standard polyethylene sheeting as a vapor barrier, retarding moisture penetration into insulated cavities.  721-02

2 **FLEXIBLE COATED FILM**
Patterned polyvinyl chloride sheets that are three-dimensional, flexible or rigid, and translucent or opaque, and in which the pattern goes all the way through the sheet. Sheets can be used as is or converted via polishing, surface embossing or printing.  1515-01

3 **WINDOW FILM**
A surface-protection system for windows and other surfaces that are vulnerable to damage. The system is made of layers of polyester film bonded by special adhesives, impregnated with UV inhibitors and finished with a patented crystal-clear, scratch-resistant coating.  1636-02

4 **HOLOGRAPHIC FILM**
A photopolymer film that records volume-phase reflection holograms. 1938-01

**5 DECORATIVE LIGHT-INTERFERENCE FILM**
The reflection and transmittal of white light from these thin (less than 0.04 in./1 mm), flexible, transparent plastic films causes light interference which the eye perceives as an iridescent color effect. 2040-02

**6 THERMOCHROMATIC FILM**
Liquid crystals are organic compounds derived from cholesterol that change color with temperature. Dispersed and encapsulated on elastic films, they can be used to detect hot spots on irregular or three-dimensional surfaces. 2794-01

**7 METALLIZED POLYMER FILM**
Body-colored, translucent scratch- and mar-resistant PVC films that produce continual repeat patterns, both vertically and horizontally. 3398-01

**8 ORIENTED FLUOROCARBON FILM**
Films that are tough, strong and clear thermoplastics. They have good chemical resistance and dielectric properties, as well as good cut and abrasion resistance and tear strength. 3230-01

4

5

6

7

8

187

**2**

**3**

**4**

### 1 FLEXIBLE PHOTOVOLTAIC FILM

Lightweight, flexible, robust, amorphous, silicon solar modules for electrical-power generation. The modules are made using a roll-to-roll manufacturing process on a polymer-web substrate 0.02 in./0.05 mm thick. 3475-01

### 2 DROPCLOTH

Wide-width, flame-retardant, non-toxic sheeting made of low-density polyethylene for all-purpose covering, screening-off and light-duty floor protection. Available static-free and UV radiation-stabilized in colors in rolls 13 ft/4 m wide. 3586-03

### 3 BIODEGRADABLE POLYMER

A biopolymer made by compounding and melting starch and other biodegradable ingredients. After it is processed into granules, it can be used for injection-molding and extruded into flat films and sheets. 3682-01

### 4 LIGHT-DEGRADABLE POLYMER

Plastic polyethylene film produced with a proprietary non-toxic additive that is approved for food packaging and can be formulated to degrade controllably when exposed to UV radiation, heat and oxygen. 4244-01

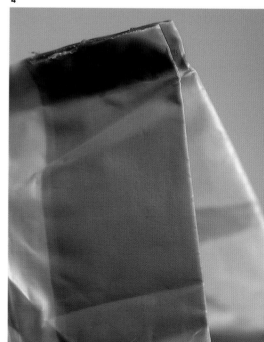

**5 THREE-DIMENSIONAL MOLDED-POLYMER LAMINATE**
This thermoformable laminate is made by pouring liquid thermoplastic film onto a temporary, flexible casting sheet. Heat-lamination then transfers the film onto a thermoformable sheet for molding into a three-dimensional part. 4570-01

**6 HEAT-WELDABLE LAMINATED SHEET**
A decorative flexible sheet in which oriented, nonwoven webbing (5 per cent polyester) is laminated onto translucent sheets of PVC. 4759-01

**7 POLYMER LACE**
Custom sheets produced by a proprietary method. Available in polyester or PVC, they can be custom-colored or laminated. Available in 45 in./114.3 cm or 34 in./86.4 cm widths with a 10 yd/9 m minimum. 4759-03

**1 ETFE ROOFING SYSTEM**
A system that uses ethylene-tetraflouroethylene sheets separated by a cushion of air secured within a frame. Air pumped into the glazing layer increases the thermal insulation without increasing weight. 4811-01

**2 PRINTING PROCESS**
A process for applying interactive graphics to flat card or board for display or promotional applications. 4816-01

**3 EDGE-BRIGHT SHEET**
By altering the wavelength of absorbed UV light (non-visible) to that of visible light, the edges of this 100 per cent polycarbonate extruded sheet give the appearance of emitting light. 4990-01

**4 LENTICULAR PRINTING PROCESS**
Lenticular printing, both digital and with a four-color offset printer, that produces a two-second (twenty-four stills) video or a three-dimensional image in a format of up to 4 by 8 ft/1.2 by 2.4 m. The image is printed directly onto glycol-modified polyethylene terephthalate (PETG) or polyvinyl chloride (PVC) instead of a medium such as paper. 4106-01

**5 ACRYLIC EDGE-BRIGHT SHEET**
Extruded plastic sheet that emits colored light. The sheet is composed of 100 per cent poly(methyl methacrylate) (PMMA) that absorbs energy from the non-visible part of the spectrum and emits it in the visible part. 5045-01

**1 HEAT-REFLECTIVE BARRIER**

A high-temperature barrier material with a metallized surface that reflects 95 per cent of radiant heat and shields against conductive heat. This sheet's five-layer structure maximizes its reflectivity and durability. 124-01

**2 ANTI-BACTERIAL UPHOLSTERY FABRIC**

A tough, durable, flame-retardant fabric made of synthetic yarns that is anti-bacterial and stain-resistant, seams nearly invisibly for wall coverings, and molds easily on furniture. Uses include seating in high-traffic areas such as airports. 10-02

**3+4+5
TECHNICAL TEXTILES**

Knitted, woven or braided textiles in a wide variety of fibers including fiberglass, carbon, aramid, polyester, polypropylene, polyamide, acrylic, metal wire and naturals. 2502-01

**6 SOLUTION-DYED ACRYLIC TEXTILE**

Inherently fire-retardant, weather-resistant and solution-dyed 100 per cent acrylic fabrics treated with a fluorocarbon and a formaldehyde-based resin. For awnings, canopies and boat tops. 2567-02

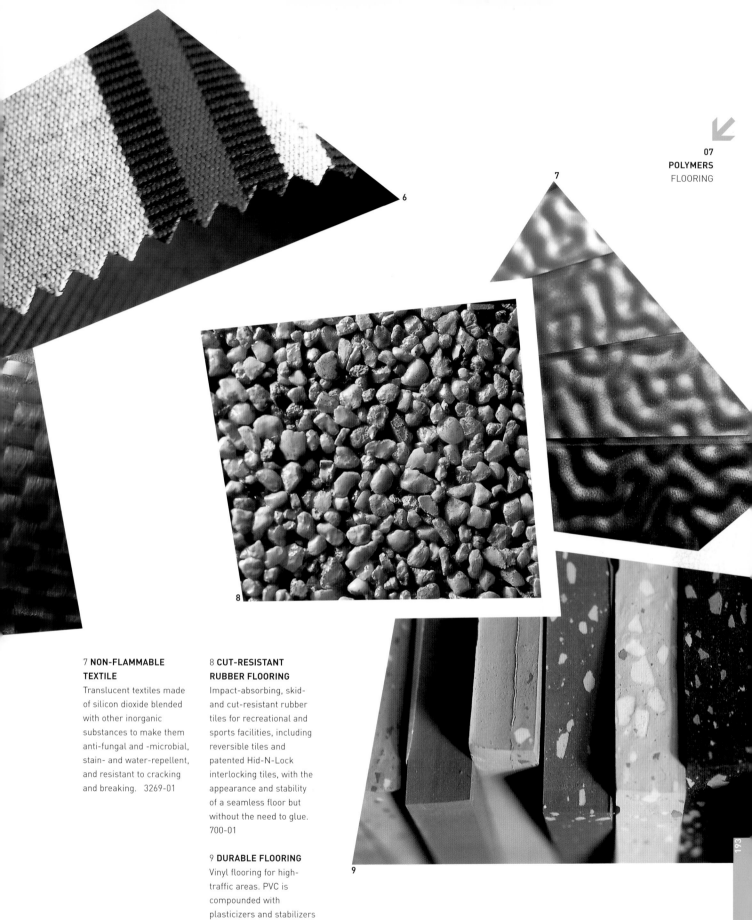

**7 NON-FLAMMABLE
TEXTILE**
Translucent textiles made
of silicon dioxide blended
with other inorganic
substances to make them
anti-fungal and -microbial,
stain- and water-repellent,
and resistant to cracking
and breaking.   3269-01

**8 CUT-RESISTANT
RUBBER FLOORING**
Impact-absorbing, skid-
and cut-resistant rubber
tiles for recreational and
sports facilities, including
reversible tiles and
patented Hid-N-Lock
interlocking tiles, with the
appearance and stability
of a seamless floor but
without the need to glue.
700-01

**9 DURABLE FLOORING**
Vinyl flooring for high-
traffic areas. PVC is
compounded with
plasticizers and stabilizers
to produce highly wear-
resistant flooring with
resistance to most acids
and alkalis as well.
Available in a wide range
of standard and custom
designs.  3585-01

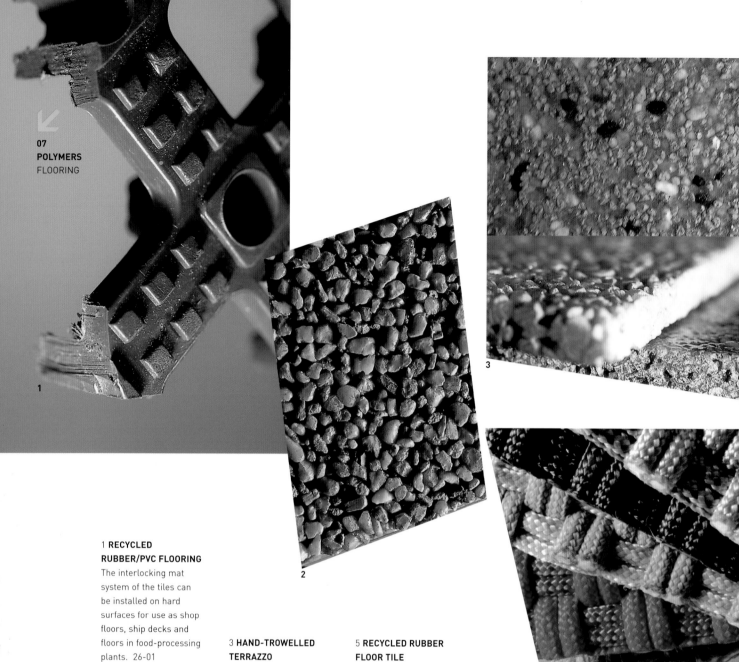

1

2

3

4

**1 RECYCLED
RUBBER/PVC FLOORING**
The interlocking mat
system of the tiles can
be installed on hard
surfaces for use as shop
floors, ship decks and
floors in food-processing
plants. 26-01

**2 HAND-TROWELLED
TERRAZZO**
A quartz/resin aggregate
that is laid as a durable
floor. A range of colors
is available to create
patterns and designs.
1528-01

**3 HAND-TROWELLED
TERRAZZO**
A 0.25 in./0.6 cm trowel-
applied chemical-, stain-,
water- and slip-resistant
monolithic decorative
flooring system. It consists
of clear 100 per cent solid
epoxy resin and colored
aggregate in almost any
color and shade. 1747-01

**4 WOVEN HIGH-
STRENGTH BRAIDED
ROPE**
High-tenacity climbing
rope composed of a
stranded nylon core and
a rubber/nylon protective
sheath woven to produce
distinctive matting and
seating surfaces. 4282-02

**5 RECYCLED RUBBER
FLOOR TILE**
A resilient surface made
from scrap-tire rubber
that provides an impact-
absorbent surface. These
safety cushions are
formed from small
particles of rubber
adhered with urethane.
4296-01

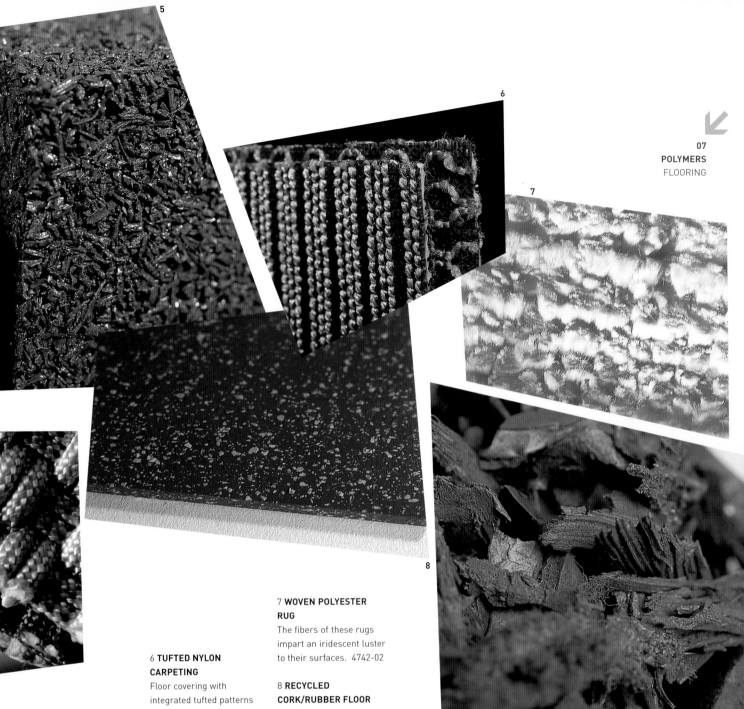

**6 TUFTED NYLON CARPETING**

Floor covering with integrated tufted patterns for interior applications. The hybrid construction provides many of the performance features of hard-surface floor coverings. Moreover, it offers slip resistance and sound-absorption and is inexpensive to maintain. 4690-01

**7 WOVEN POLYESTER RUG**

The fibers of these rugs impart an iridescent luster to their surfaces. 4742-02

**8 RECYCLED CORK/RUBBER FLOOR TILE**

Composite tiles composed of 60 per cent recycled cork and 40 per cent recycled rubber. They are durable, water-resistant and available in seventeen different fade-resistant colors and two finishes. 4571-01

**9 RECYCLED RUBBER GROUND COVER**

Loose-fill ground cover derived from 100 per cent recycled vulcanized rubber from passenger and/or truck tires. The coloring process molecularly bonds non-toxic colorants to shredded, recycled tire grinds. 4789-01

1

2

3

### 1 FLEXIBLE POLYMER MOLDING

Available as preformed 'Made to Fit' polyester molding – custom-manufactured to extremely tight radii according to blueprint specifications or template drawings – or as 'Bend to Fit' polyurethane molding.  191-02

### 2 RIGID FOAMS

High-density polyurethane foams that sense the body's weight and temperature and respond by molding to its exact shape and position, thus distributing the weight and reducing stress on pressure points.  4191-02

### 3 VIBRATION-DAMPING FOAMS

The surfaces of these molded-elastomer vibration- and shock-isolation pads have recessed, offset cells. This construction allows the elastomer to flow under load while maintaining lateral stability, thereby providing positive grip to the machine foot and the floor when under vibration. 169-01

### 4 ACOUSTIC-DAMPENING COMPOSITE SHEET

A thin, acoustically formulated composite sheet-membrane that reduces impact and airborne sound. Composed of nonwoven polyester laminated to both sides of a chlorinated polyethylene, it is used for acoustical treatments over common substrates.  1729-01

### 5 FOAM PACKING

Flexible or rigid packaging material made of polyurethane foam in a range of densities and in various colors. The flexible foam is available in both molded and sheet form, die- or square-cut, and in simple or complex configurations. Both kinds of foam come in a range of densities and colors. 100-04

## 6 STEELCASE
### CACHET CHAIR
A stackable chair that utilizes elastomer springs to create suspension. Thermoplastic elastomer (TPE) connections at key points allow the seat back to rock, giving it independent motion and responding to each sitter's comfort needs.

## 7 PAOLA LENTI
### SAND LOUNGE CHAIR
A fixed-position chair for outdoors. The frame is made of galvanized cataphoresis-treated steel and polyester varnish; the feet are made of brushed stainless steel. The covering is composed of waterproof, non-toxic, hypoallergenic, anti-mold Rope fabric with an excellent colorfastness to UV rays, chlorine and seawater.

6

4

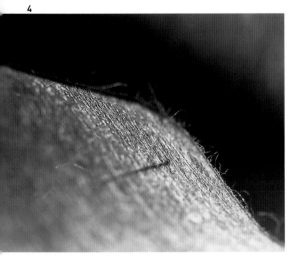

5

7

**1 FLEXIBLE IMPACT-ABSORBING FOAM**
A soft, high-density polyurethane foam pad that contours to the body and transpires moisture away from it. It absorbs jolts and vibrations. 126-02

**2 STRUCTURAL SANDWICH PANEL**
A high strength-to-weight-ratio sandwich material composed of a structural cellular plastic core with facings made of fiber-reinforced plastic. Applications include marine construction, pipelines, trains and aerospace. 128-01

**3 RETICULATED FOAM**
Flexible polyurethane foam with controlled cell size. Its tensile strength, tear strength and capacity for elongation are two or three times higher than those of conventional polyurethane foams. It can be fabricated using a wide range of techniques. 159-02

**4 RECYCLED INSULATION**
Recycled polyurethane makes up this sound insulation for floor, ceiling or wall. 1918-01

**5 HIGH-DENSITY ENERGY-ABSORBING FOAM**
Microcellular, open-celled urethane foams with high energy absorption. They are resistant to collapse and chemicals, and withstand short-term exposure to temperatures up to 250ºF/121ºC. 2633-01

**LUMINEX SHIRT**
A shirt that incorporates woven optical fibers for decorative effect. Thin optical fibers are woven into a synthetic fiber, the ends of which are bundled together to a point from which LED light is transmitted through the filaments. These emit light along their length and also from the filament ends.

**6 FINE-CELL FOAM TAPE**
Tape-grade foam with a
smooth surface. It is
non-toxic, flexible, elastic,
resilient, rot- and UV-
resistant, and complies
with FDA regulations for
food contact. Designed
primarily for adhesive-
based applications.
1994-04

4

5

6

**1 CONFORMABLE MEDICAL-GRADE FOAM**
Odorless, molded polyurethane foam for medical products such as cushions, pillows and mattresses, it conforms to the shape of the body and returns to its original form after use. Available in three color-coded densities. 3652-01

**2 OPEN-CELL FOAM INSULATION PANEL**
Thin, efficient panels consisting of a polystyrene-foam core encapsulated in a sealed and evacuated barrier material such as a metal foil or metallized film laminate. 3480-01

**3 STRUCTURAL SANDWICH PANEL**
Strong, structural composite sheets designed for use in the marine industry. Flexible composites with a continuous-use temperature of 200ºF/93ºC that are rot- and chemical-resistant and bond to a wide range of materials. 4192-01

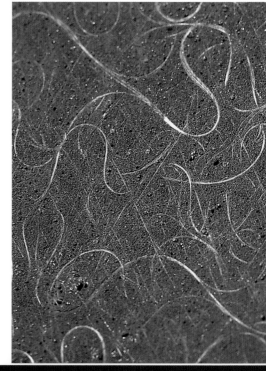

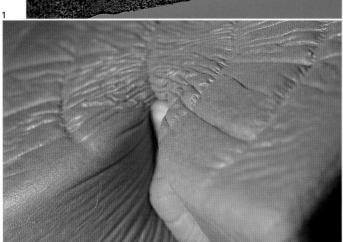

4

**4 INSULATED DRAINAGE PANEL**
Geo-synthetic, insulated panels composed of molded blocks of expanded polystyrene beads bound with a waterproof binder on a laminated filter backing.
4590-01

**DAKOTA JACKSON**
DB-1
An office table that comprises a glass/acrylic sandwich work surface with a skin-like PVC interlayer. The fluorescent acrylic sheet provides an ethereal glow underneath a glass top surface. The vinyl sheet is a continuous loop that travels between the glass and acrylic sheets as well as beneath the tabletop.

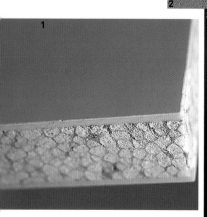

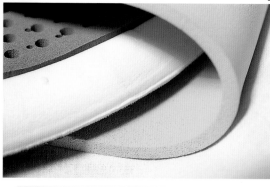

### 1 LIGHTWEIGHT COMPOSITE PANEL

It is weatherproof, insulating and chlorofluorohydrocarbon (CFC)-free, and is composed of two cover sheets of glass fiber-reinforced polyester laminates about 0.06 in./1.5 mm thick with a white semi-matte finish. 1433-01

### 2 RIGID PVC FOAM PANEL

Foamed, closed-cell PVC with a high strength-to-weight ratio that can be fabricated using standard tools and screen-printed without priming. Available in a range of thicknesses. 1627-01

### 3 POLYOLEFIN FOAM ROLL

A recyclable foam for sound absorption composed of 100 per cent polyolefin foam made from close-packed pellets and expanded using a foaming agent. The roll can be formed to any shape. 4117-02

### 4 VIBRATION-DAMPING FOAM

Flexible, soft, anti-shock thermoplastic foam for applications with low transmitted force. It absorbs up to 94 per cent of impact energy and does not collapse after impact, though a severe impact may reduce its performance by 10 to 15 per cent. 4739-03

### 5 FOAM-CORE RIGID PANEL

Large, flat, lightweight plastic panels with a foamed inner core. They have good sound-dampening and insulating properties, are moisture-proof and mold-resistant, and can be worked like wood. 89-04

**6 HIGHLY ELASTIC CUSHIONING MATERIAL**
A unique type of cushioning made of highly elastomeric materials in a column construction. A wide variety of properties can be engineered by changing the formulation of the elastomer and the dimensions of the column walls and cells. 4056-02

**7 FLUID FOAM**
A lightweight foam composed of acrylic microspheres (less than 0.008 in./200 μ in diameter) and proprietary viscoelastic lubricants that facilitate sliding and rolling between the microspheres. 4056-03

**8 CUSHIONING GEL**
A fatigue-resistant (survives hundreds of thousands of deflections), thermoplastic, elastomeric gel with texture that can be controlled from stiff to extremely soft by varying its oil content. 4056-04

**9 FLUID FOAM**
Very lightweight materials that are impact-absorbing and cushioning. These non-flammable, non-toxic composites, constructed of microspheres plus a bonding agent, are encased in a polyurethane or polyvinyl chloride film and conform to shape under stress. 4242-01

6

7

8

9
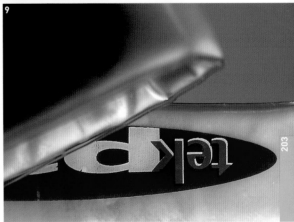

203

07
**POLYMERS**
GELS +
HIGH-PERFORMANCE TEXTILES

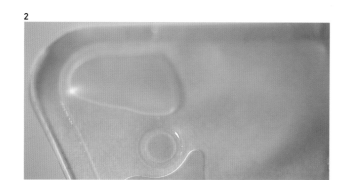
2

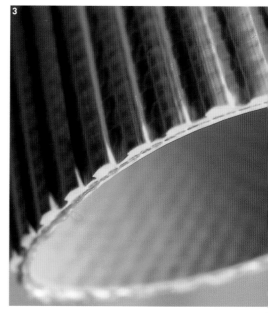
3

1

4

**1 CUSHIONING GEL**
A breathable, dimensionally stable, polyurethane-based gel that is shock-absorbant, flexible and elastic, and distributes pressure equally. It contains no plasticizers or other volatile components.
1524-01

**2 LIQUID-ENCAPSULATED CUSHIONING**
An advanced padding technology that uses liquid to cushion the body from shear force, pressure and shock, thereby reducing pain and discomfort.
4592-01

**3 LIGHTWEIGHT BRAIDED HOSE**
Flexible hoses that are abrasion-, heat-, chemical-, oil- and weather-resistant. They are made of a technical textile of woven mixed polyester and polyamide yarns, lined with a nitrile elastomer and covered with a ribbed nitrile elastomer.  3800-01

**4 SELF-SUPPORTING WOVEN TEXTILE**
An elastomeric, openwork upholstery fabric woven (by a proprietary process) of DuPont Hytrel®, a flexible polyester elastomer, as the major component plus other fibers. It is stable, breathable and conformable, and combines the functions of support and cushioning.
3448-01

8

## 5 PROTECTIVE BRAIDED SLEEVING

Expandable monofilament sleevings used for protective coverings. A wide range of polymer fibers (e.g. polypropylene, polyester, fiberglass, nylon, fluoropolymer, polyphenylene sulfide, depending on the application) are fabricated using three-dimensional braiding.
4104-01

## 6 SHOELACE

Braided cords designed for use as shoelaces. The patent-pending collapsible knot is tension-adjusted between each set of eyelets. When the tension is released, the knot reappears, thus securing the laces in place for a comfortable fit. 4398-01

## 7 THREE-DIMENSIONAL SPACER FABRIC

Three-dimensional fabrics that are warp-knit on a Raschel machine in a single knitting sequence, which produces two face fabrics independently constructed and connected by 'spacer yarns'. 4258-01

## 8 THREE-DIMENSIONAL SPACER FABRIC

Raschel-knit, 100 per cent polyester upholstery textiles that have a depth and cushioning similar to polyurethane foam. Offered in a wide range of colors. 4870-01

## BUILT NY

BYOBAG
Neoprene is most commonly used for wetsuits: it insulates, cushions and stretches. More recently these properties have been harnessed to protect and maintain the temperature of wine bottles.

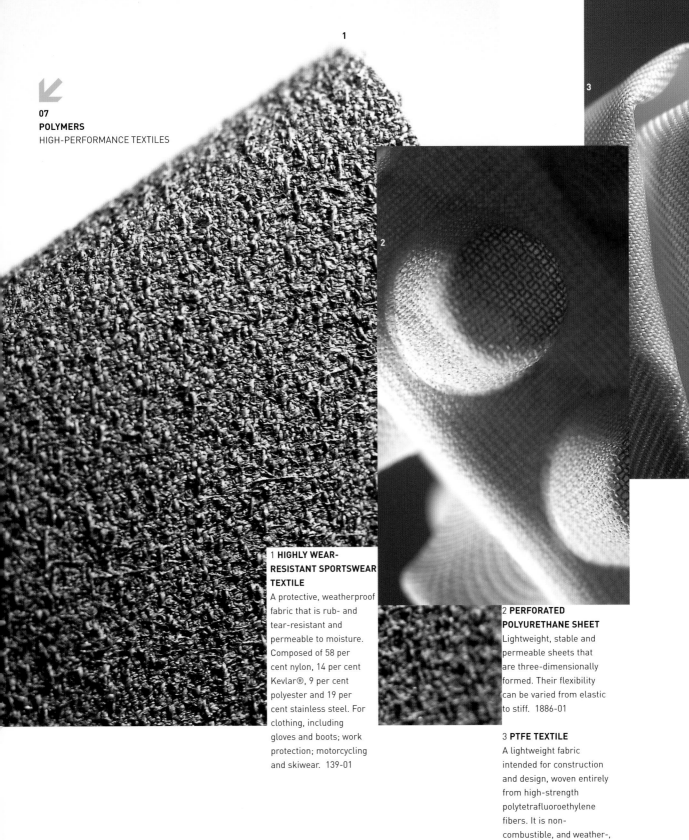

**1 HIGHLY WEAR-RESISTANT SPORTSWEAR TEXTILE**
A protective, weatherproof fabric that is rub- and tear-resistant and permeable to moisture. Composed of 58 per cent nylon, 14 per cent Kevlar®, 9 per cent polyester and 19 per cent stainless steel. For clothing, including gloves and boots; work protection; motorcycling and skiwear. 139-01

**2 PERFORATED POLYURETHANE SHEET**
Lightweight, stable and permeable sheets that are three-dimensionally formed. Their flexibility can be varied from elastic to stiff. 1886-01

**3 PTFE TEXTILE**
A lightweight fabric intended for construction and design, woven entirely from high-strength polytetrafluoroethylene fibers. It is non-combustible, and weather-, chemical- and UV-resistant. 2277-01

4

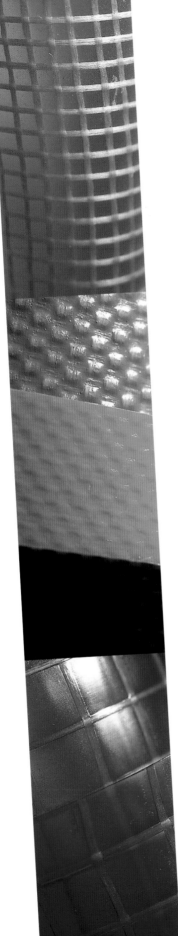

5

6

7

**4 AIR- AND WATERTIGHT DOUBLE-WALL FABRIC**
The walls of this drop-stitch, woven-polyester fabric are separated by 2.68 in./67 mm. It is coated with polyvinyl chloride, is tear-resistant and has a high tensile strength. 2318-01

**5 COATED INDUSTRIAL FABRIC**
Woven polyester fabrics that are available with the following coatings: polyvinyl chloride, thermoplastic polyurethane, thermoplastic polyester, polypropylene and Alcryn® (melt-processible rubber). 2318-02

**6 COMPOSITE FORMING PROCESS**
A fabricating technique for shaping and forming a composite film on pre-shaped molds so that the fabric is essentially seamless and thus less prone to stretch. In this case, the film is made from Mylar® reinforced with aramid yarn. 3474-01

**7 BALLISTIC TEXTILE**
Various nonwoven materials made by a patented process in which parallel strands of synthetic fibers (Spectra® polyethylene or high-strength aramid fibers) are laid side by side and held in place by thermoplastic, thermoset or blended resins. 3477-02

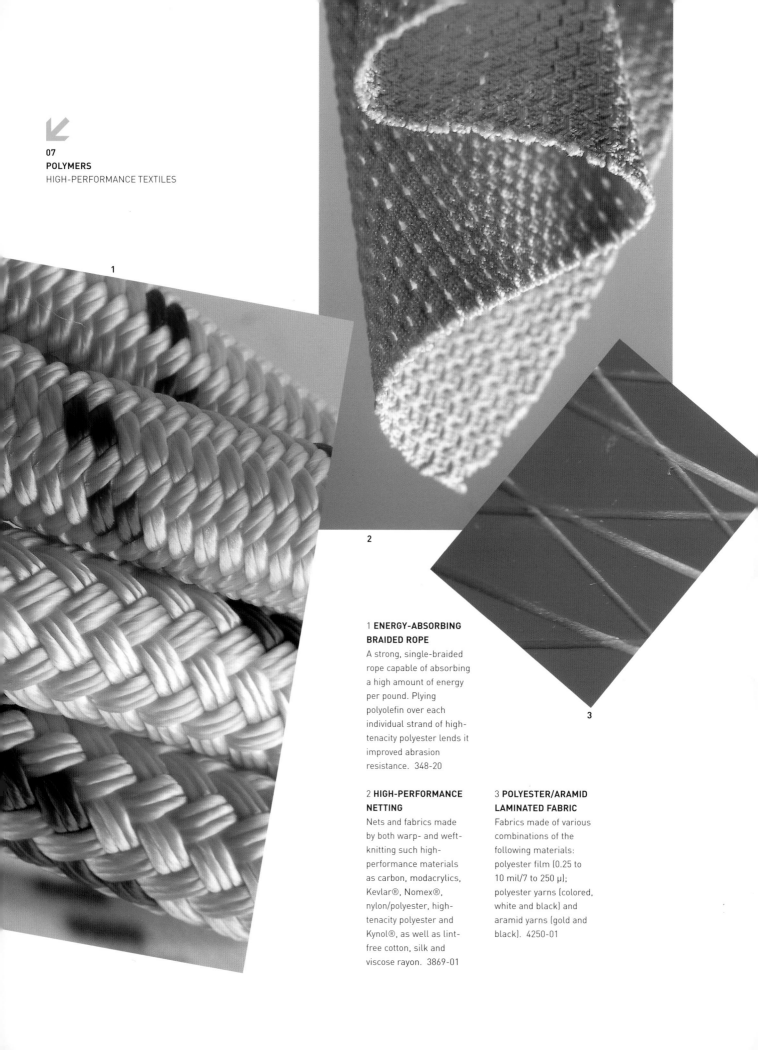

1

2

3

**1 ENERGY-ABSORBING
BRAIDED ROPE**
A strong, single-braided
rope capable of absorbing
a high amount of energy
per pound. Plying
polyolefin over each
individual strand of high-
tenacity polyester lends it
improved abrasion
resistance.  348-20

**2 HIGH-PERFORMANCE
NETTING**
Nets and fabrics made
by both warp- and weft-
knitting such high-
performance materials
as carbon, modacrylics,
Kevlar®, Nomex®,
nylon/polyester, high-
tenacity polyester and
Kynol®, as well as lint-
free cotton, silk and
viscose rayon.  3869-01

**3 POLYESTER/ARAMID
LAMINATED FABRIC**
Fabrics made of various
combinations of the
following materials:
polyester film (0.25 to
10 mil/7 to 250 μ);
polyester yarns (colored,
white and black) and
aramid yarns (gold and
black).  4250-01

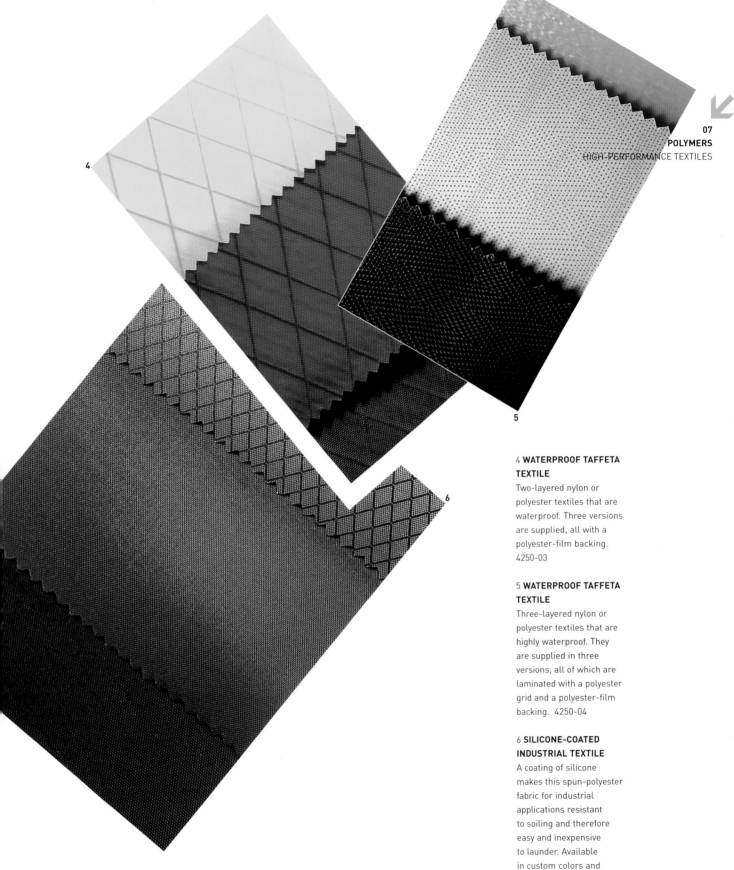

4 **WATERPROOF TAFFETA TEXTILE**
Two-layered nylon or polyester textiles that are waterproof. Three versions are supplied, all with a polyester-film backing. 4250-03

5 **WATERPROOF TAFFETA TEXTILE**
Three-layered nylon or polyester textiles that are highly waterproof. They are supplied in three versions, all of which are laminated with a polyester grid and a polyester-film backing. 4250-04

6 **SILICONE-COATED INDUSTRIAL TEXTILE**
A coating of silicone makes this spun-polyester fabric for industrial applications resistant to soiling and therefore easy and inexpensive to launder. Available in custom colors and finishes. 4250-09

2

3

1

5

4

6

### 1 HIGH-TENACITY KNITTED CORD

Two types of plastic cord that have high tensile strength. They are resistant to abrasion, weather, mold, rot and mildew. 4403-01

### 2 LAMINATED POLYAMIDE NETTING

A netting formed by PVC-laminated polyamide sheets cut with curved slits to produce a leaf effect for camouflage. The mesh is then separately sprayed on each side in a range of seasonal shades depicting summer, autumn, snow or desert. 4400-01

### 3 FLEXIBLE-FIBER COMPOSITE-FORMING PROCESS

A patented process for manufacturing flexible fiber-reinforced polyurethane composites with controllable breathability in which the fiber-polyurethane bond is made deliberately weak, thereby allowing the fibers to move within the matrix and to maximize flexibility of the composite. 140-01

### 4 WIND-SHIELDING TEXTILE

A breathable barrier that stops 80 to 90 per cent of wind penetration into a knit garment while being three to four times more breathable than typical film-core laminates. 3749-01

### 5 OMNI-DIRECTIONAL STRETCH FABRIC

An elastic material consisting of a two- or three-ply Lycra® spandex (20 per cent) and nylon (80 per cent), or Lycra® spandex (9 per cent), polyester (51 per cent) and nylon (40 per cent) warp-knit fabric laminated with a polyurethane film. 3849-01

### 6 LIGHT-INTERFERENCE COATED TEXTILE

Textiles that contain light-interference pigments incorporated into cast polyurethane resin and laminated to circular-knit polyester fabrics. The light-interference pigments create very saturated, bright colors that shift with the angle of vision. 2604-03

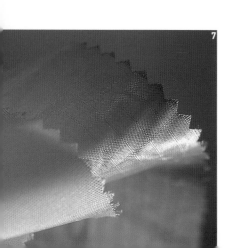

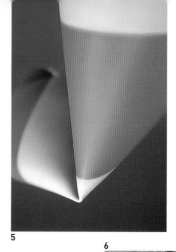

**1 THERMOCHROMATIC-COATING TEXTILE**
Changes in temperature make this flexible thermoplastic sheet change color. Composed of polyurethane coated with a liquid-crystal emulsion and laminated to a polyester-knit backing. 2604-04

**2 HOLOGRAPHIC PRINTED TEXTILE**
Fabrics that have three-dimensional holographic designs on their surfaces. These are two-way-stretch, drapeable and washable materials composed of a diffraction-grating holographic transfer foil laminated to a polyurethane face. 2604-06

**3 IRIDESCENT UPHOLSTERY TEXTILE**
Fabrics composed of 77 per cent polyurethane fabric with a 23 per cent polyester double-knit backing. The fabric is available in twenty different colors derived from aerospace-industry dichroic pigments. 2604-09

**4 COATED BREATHABLE SPORTSWEAR TEXTILE**
Nylon and nylon/polyester textiles that have breathable coatings of nylon and Teflon®; available in all colors. 3434-01

**5 PHOTOLUMINESCENT TEXTILE**
Vinyl-coated cotton-polyester fabrics. The flame-retardant PVC comes in three colors: photoluminescent yellow, fluorescent lime and photoluminescent lime. 4194-03

**6 EMBOSSED UPHOLSTERY FABRIC**
This wide range of debossed and embossed upholstery fabrics is made from materials such as acrylic, polypropylene, polyester, wool and Trevira® CS. 4866-01

**7 FIRE-RETARDANT SAILCLOTH**
100 per cent-woven high tensile-strength sailcloth with high warp and weft tear-resistance. This cloth passes the BS5867 (British Standard) fire rating, self-extinguishing during face ignition. 4514-01

**NISSAN**
SUNCUBE
Translucent, lightweight polyester sheets can be used to sandwich an unlimited range of different materials, in this case beach grass. Because they are inexpensive and widely available, and can be cut and shaped easily, they are ideal for the consumer market.

**4 HDPE-MONOFILAMENT SHADE CLOTH**
Reflective, weather-resistant and flame-retardant screens made of UV radiation-stabilized high-density polyethylene (HDPE) monofilaments and 100 per cent pure aluminum. Available in various styles for protecting crops from solar heating. 3472-01

**3 EXTRUDED POLYPROPYLENE NETTING**
Plastic netting made primarily of continuously extruded polypropylene in thicknesses from 0.01 to 0.25 in./0.04 to 0.64 cm), with hole sizes from 0.03 to 4 in./0.06 to 10.2 cm). It is thermoformable so can be sealed to itself. 3411-01

**1 OUTDOOR WOVEN POLYESTER FABRIC**
Vinyl-coated fabrics with good tensile and tear strength that are flame-, abrasion- and mildew-resistant, colorfast and washable; applications include cushions and umbrellas for outdoor furniture, banners and structures. 1494-05

**2 AGRICULTURAL SHADE CLOTH**
A group of woven synthetic fabrics (e.g. polyethylene, polypropylene, polyester, UV-stabilized polyolefin) used for various purposes such as agricultural shade cloth, poultry and livestock curtains, insect-control screening and ground cover. 2060-01

5 **POLYESTER BRAIDED TAPES**
Strong, flexible tapes of polyester, cotton and/or rayon in various proportions in colors that give the tapes a 'wood' look; for handbags, belting, drapery and pillow trim, and accessories. 3530-01

6 **HDPE RASCHEL-KNIT MESH**
A fabric made of 100 per cent high-density polyethylene (HDPE). Available in black, white, green, brown, tan, smoke, blue, jade and custom colors, and as a fire-retardant fabric; for applications such as shade cloth and truck covers. 4437-01

7 **POLYPROPYLENE MESH**
Netting made of high-tenacity polypropylene or polyester, depending on the application, in various mesh sizes and thicknesses, in black, green, red and blue. Uses include container covers, load-carrying, circular slings, and safety and catch nets. 3876-01

5

6

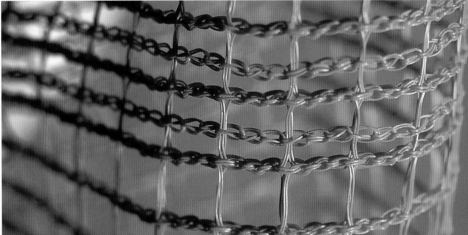

7

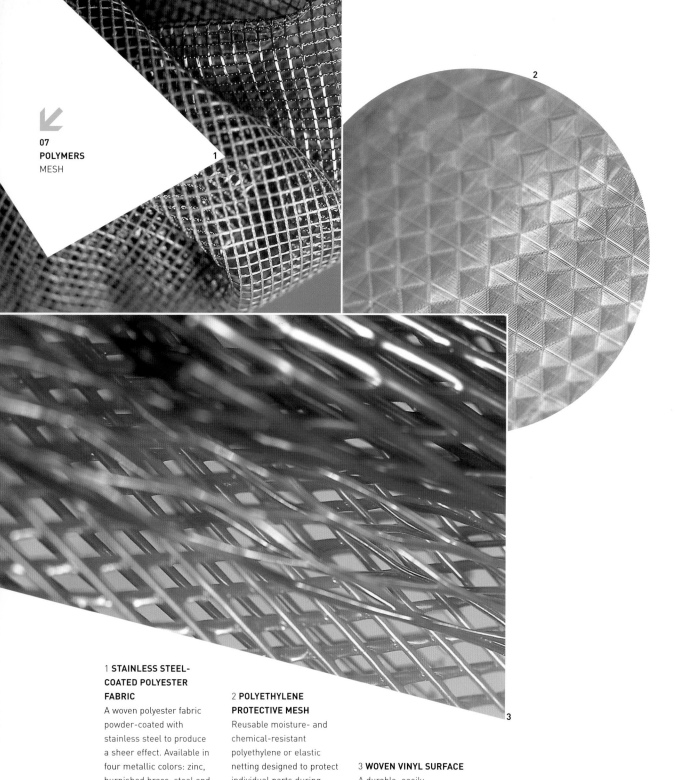

1

2

3

1 **STAINLESS STEEL-
COATED POLYESTER
FABRIC**
A woven polyester fabric
powder-coated with
stainless steel to produce
a sheer effect. Available in
four metallic colors: zinc,
burnished brass, steel and
titanium.  289-01

2 **POLYETHYLENE
PROTECTIVE MESH**
Reusable moisture- and
chemical-resistant
polyethylene or elastic
netting designed to protect
individual parts during
handling, storage and
transportation. Available
in four mesh types and
colors.  116-01

3 **WOVEN VINYL SURFACE**
A durable, easily
cleanable, fade-resistant
woven vinyl covering
designed to be non-
slippery even when wet.
It can be installed indoors
or outdoors.  440-01

5

4

**4 POLYPROPYLENE MIST-ELIMINATING MESH**
Plastic mesh manufactured in polypropylene and designed to separate liquid from vapor down to 1 μ. The ladder arrangement of the two sets of filaments causes a change in the direction of vapor flow that enhances droplet removal.  1418-01

6

**5 MONOFILAMENT GEO-TEXTILE**
A patented, open, three-dimensional woven geo-textile that is strong and durable, composed of UV-stabilized polypropylene monofilament yarns woven into a dimensionally stable uniform configuration. 2060-02

**6 MULTILAYER GEOCOMPOSITE**
A lightweight geocomposite composed of a nonwoven geo-textile heat-bonded to a durable, open, three-dimensional polymer core. The geo-textile holds back the adjacent soil and helps develop a natural soil filter while allowing water to seep into the core. 1426-01

**1 INSECT NETTING**
Nets woven of acrylic and high-density polyethylene (HDPE) yarns that have been stabilized. Six different types are available with different hole sizes, some with rectangular instead of square openings. 3472-02

**2 PTFE-COATED SCREEN**
Made of polyester or Teflon®-coated polyester, these strong, heavy woven and spiral screens have a continuous service temperature of 347°F/175°C. They are fabricated in standard sizes up to 23 ft/7 m wide and 164 ft/50 m long. 4246-01

**3 RIGID THREE-DIMENSIONAL MESH**
A transparent, molded polycarbonate mesh produced to eliminate moisture build-up in insulation blankets. It is lightweight, flame-resistant, crushproof and flexible and can be cut with scissors. 117-03

**4 NONWOVEN MESH**
Nylon mesh that can be compressed to create different surface textures. Applications include set and interior design. 4759-05

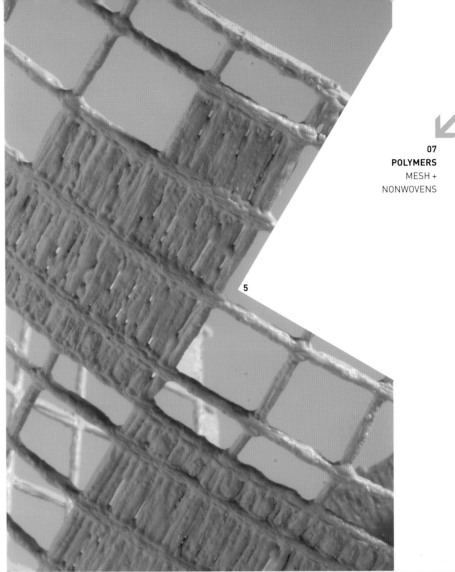

5

4

**5 KNITTED SHADE CLOTH**
A material with a triangular pattern. This patented geometry of net design reduces heat accumulation by up to 18 per cent underneath the shade by creating constantly moving shadowing as the earth rotates. 4965-01

**6 WEB ADHESIVE**
Adhesives for high-performance textile-laminating applications. These adhesives incorporate hot-melt adhesives (they are solid at room temperature, liquefy when heated and re-solidify when cooled) with melt-blown fiber webs formed using modified polyester, polyamide or polyolefin fibers. 253-02

**7 PERFORATED FELT**
Felt made of 100 per cent polypropylene, 100 per cent polyester or a blend in any percentage that does not contain any binding agents and is colored by the addition of pigments. Breathability can be adjusted by the size, placement and number of perforations. 3785-01

6

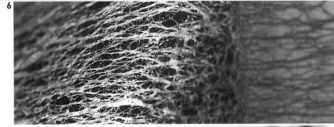

7

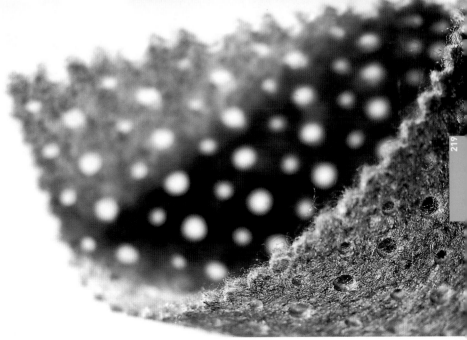

### 1 *WASHI* PAPER

Flame-resistant, rice paper-like fabric made of a cellulosic/fiberglass nonwoven and intended for exhibits, displays and building environments as wall screens, digital and screen prints, printed panels, tubes and banners. 1428-01

### 2 HDPE PRINTING SUBSTRATE

A substrate made of spun-bonded high-density polyethylene (HDPE). This material combines some of the properties of paper, film and cloth. It is lightweight, dimensionally stable and opaque, and has a smooth surface. 1930-01

### 3 POLYPROPYLENE FABRIC

A nonwoven, this strong, dimensionally stable fabric is made by spinning and thermally bonding continuous filaments of polypropylene in a directionally oriented configuration. It has high tensile and tear strength. 2490-01

### 4 FOOD-SAFE POLYPROPYLENE NONWOVEN TEXTILE

Spunbonded textiles of 100 per cent polypropylene that have a soft hand, are safe for food and beverage contact, and are custom-designed to be hydrophilic or hydrophobic, flame-retardant, antistatic, UV-resistant and/or antibacterial. 2493-08

### 5 *FAUX* DOESKIN FABRIC

A synthetic textile with the appearance and feel of doeskin suede. Composed of nonwoven, 100 per cent polyester ultra-microfibers set in a polyurethane binder, it is durable, resilient, crock- and pilling-resistant, dry-cleanable and machine-washable. 2841-01

### 6 SPUN-BONDED POLYESTER FABRIC

A sheet with a smooth surface that contains no resins, sizing or binders. It has high tensile and tear strength, non-raveling edges, controlled permeability, good chemical resistance and dimensional stability. 3086-01

### 7 POLYESTER NONWOVEN FABRIC

A textile made of staple polyester fiber that is printed with a high-friction surface of rubber dots on one side and a low-melt adhesive on the other side. When fused to rugs and carpets, this surface prevents slippage. 3136-02

### 8 FIRE-RETARDANT CELLULOSE/GLASS FIBER

A thin, homogeneous, acoustic nonwoven textile made of cellulose and glass fibers, which is glued to the back of perforated ceiling tiles (of metal, wood, plastic or gypsum) by means of an adhesive that is already applied to it. 3136-01

### 9 *FAUX* LEATHER

A luxury leather-alternative fabric that is durable, colorfast and stain- and abrasion-resistant. Composed of a 100 per cent polyurethane top surface laminated onto 100 per cent rayon backing, the fabric simulates the look and feel of real leather. 3687-01

### 10 PRESSED FELT MEMBRANE

A woven, nylon-base membrane is strengthened and made thicker by needling several layers of nonwoven nylon bats onto it. The pressed felts are used to dry paper. 4246-02

### 11 FELT

These engineered, wet-processed, machined wool-and-polymer felts for industrial applications are produced in thicknesses from 0.06 to 1 in./0.15 to 2.54 cm, and in densities from 0.1 to 0.2 oz/in.$^3$/0.18 to 0.34 g/cm$^3$. 4492-01

**TORD BOONTJE**
MIDSUMMER LIGHT
The delicate, intricate floral designs of Midsummer Light are laser-cut from Tyvek®, a heat-resistant, spun-bonded, non-woven polymer paper. The flowers can be draped around the lamp in a pattern pleasing to the installer.

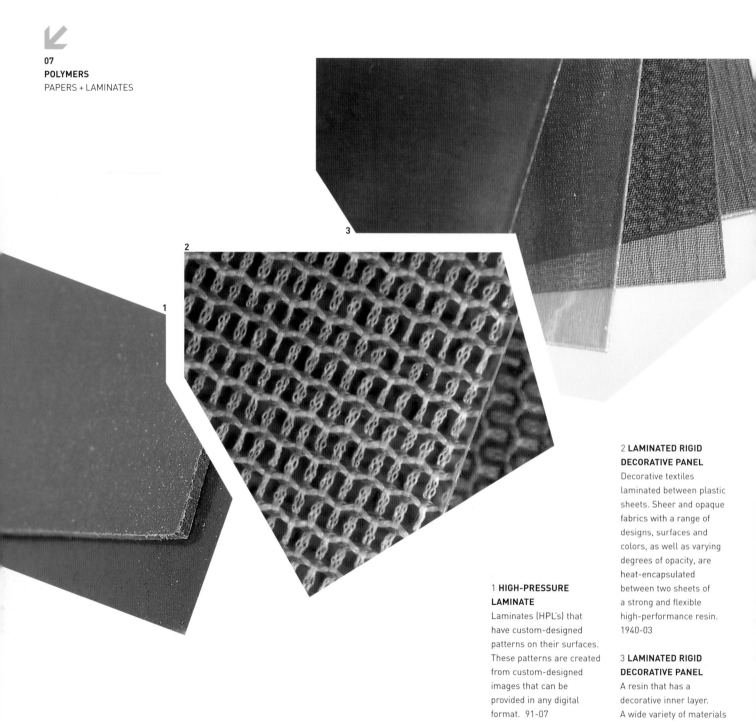

**2 LAMINATED RIGID
DECORATIVE PANEL**
Decorative textiles
laminated between plastic
sheets. Sheer and opaque
fabrics with a range of
designs, surfaces and
colors, as well as varying
degrees of opacity, are
heat-encapsulated
between two sheets of
a strong and flexible
high-performance resin.
1940-03

**3 LAMINATED RIGID
DECORATIVE PANEL**
A resin that has a
decorative inner layer.
A wide variety of materials
such as natural fibers,
textiles, films and graphic
prints are laminated
between acrylic sheets.
4430-01

**1 HIGH-PRESSURE
LAMINATE**
Laminates (HPL's) that
have custom-designed
patterns on their surfaces.
These patterns are created
from custom-designed
images that can be
provided in any digital
format.  91-07

**4 TRANSLUCENT DECORATIVE SHEET**

Laminate sheets with three-dimensional decorations. Created with translucent paper and melamine resin, they are available in five colors and twenty-one different patterns. 91-02

**5 DIGITALLY PRINTED HIGH-PRESSURE LAMINATE**

Custom-designed digital images create the patterns on these laminates. 91-07

**6 ELECTRICAL-GRADE ARAMID PAPER**

This paper is corrugated to produce transformer-winding insulation. Square-, rectangular- and sinusoidal-shaped corrugations are available in a range of depths. 152-02

**7 DIMENSIONALLY STABLE EXTRUDED POLYMER**

A durable, acrylic/PVC alloy that is dimensionally stable, flame-retardant, and impact- and chemical-resistant. It is extruded in a range of colors, patterns, textures, thicknesses and grades, and is machinable using conventional power tools. 1226-01

**8 CURVED HIGH-PRESSURE LAMINATE**

Duro-plastic laminates that have high abrasion resistance and impact strength. The sheets are available up to 161.4 by 51.2 in./410 by 130 cm in gloss, matte and hammer-embossed, fine-grain and soft finishes. 1825-01

4

5

6

7

8

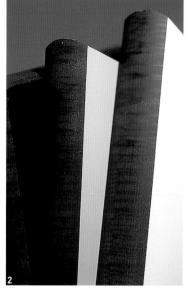

**1 TEAR-RESISTANT POLYMER PAPER**
A recyclable synthetic paper that is durable, strong, waterproof, smooth-surfaced and chemical-, grease- and tear-resistant. It is composed of polypropylene resins, inorganic fillers and additives, and has a multi-layer construction. 2771-01

**2 IMPACT-RESISTANT PHENOLIC SOLID SURFACING**
To make these composite panels, high temperature and pressure are used to set a phenolic, resin-based thermoset. Used as chemical-, corrosion- and impact-resistant work surfaces. 4380-01

**3 DICHROIC GLASS/POLYMER PANEL**
Laminate and glass panels that exhibit a full-spectrum color shift depending on the viewing angle. Made of polycarbonate or glass, a proprietary polymer and vinyl or phenolic polymers or paper. 3797-01

**4 ELECTRICAL-GRADE ARAMID PAPER**
Nomex® paper, a mechanically strong and flexible material that has good electrical properties at high temperatures and is produced from two forms of an aramid polymer. 4499-01

**5 DECORATIVE LAMINATE SHEET**
Translucent resin and paper create a decorative effect in this 100 per cent cast, three-ply polyurethane laminate. The top coat, used to protect the surface, is available in matte, gloss or textured and may also be made of hard or soft resin. 4817-01

**6 DECORATIVE LAMINATE SHEET**
As with the previous example, translucent resin and resin dust create a decorative effect in this 100 per cent cast, three-ply polyurethane laminate. 4817-02

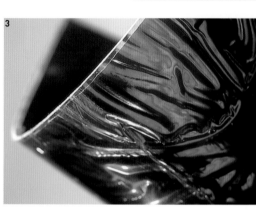

**7 LAMINATED ACRYLIC SHEET**
Custom acrylic-laminated rice paper composed of a layer of paper between two layers of Acrylite®. It may be manufactured in sizes up to 5 by 10 ft/ 1.52 by 3.05 m. 4601-01

**8 TEAR-RESISTANT POLYMER PAPER**
Highly durable synthetic printing film composed of high-density polyethylene formulated to accept high-resolution print from a wide range of printing types. 4818-02

**9 DECORATIVE LAMINATE SHEET**
Another example in which translucent resin creates a decorative effect in a 100 per cent cast, three-ply polyurethane laminate. 4817-03

**10 WATERPROOF BOOK-MAKING PROCESS**
In this proprietary binding process, polymer-derived paper is printed with colorfast ink to produce a durable, 100 per cent waterproof book which will not fade or deteriorate. 4970-01

**11 WEATHER-RESISTANT EXTRUDED PANEL**
Polycarbonate film and a proprietary thermoformable, weather-resistant film are co-extruded and can be applied to metal surfaces. This film can replace painted panels in automotive applications. 4921-01

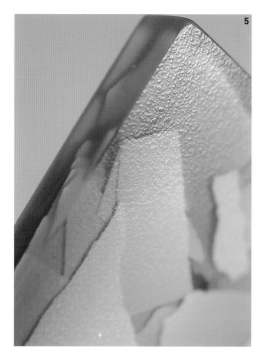

5

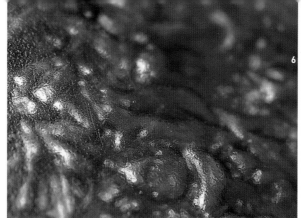

6

8

aleron
stren...h films

WIRE PULL SAMPLE
TRY TO PULL IT
1-713-465-6111

9

7

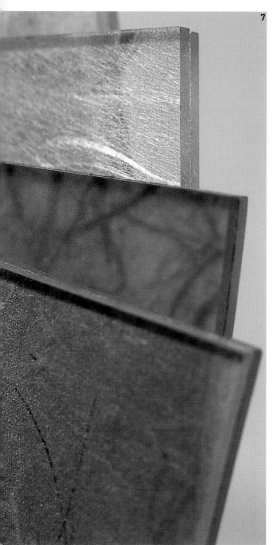

10

11

1

2

**1 DECORATIVE POLYMER BEADS**

Smooth colored beads used to decorate surfaces. They are made of 100 per cent PVC that can be adhered to plastic or fabrics with adhesive and can be applied by laminating, screen-printing or stenciling. 3901-02

**2 PHOTOLUMINESCENT PAINT**

A polyurethane paint that is chemical-resistant and highly elastic. It contains non-toxic, non-radioactive zinc sulfide crystals (pigment) encased in flexible or rigid sheets. 4194-03

**3 WATERSLIDE DECAL PAPER**

A range of transfer papers for decoration. Available in three types made of specially formulated twin-wire paper for decorating pottery, glass, vitreous enamelware and precious metals, and for heavy enamel effects. 4241-01

**4 PHOTOCHROMIC INK**

Ink colors that appear outdoors and disappear indoors. These screen-printing inks, which are available in eighteen colors plus custom colors, are non-toxic and environmentally safe, do not dry on the screen and contain no solvents. 4326-01

**5 SPRAY-DEPOSITION PROCESS**

A line-of-sight process for painting small polymer and glass products using robot manipulation. Any geometry of surface may be covered. This process coats plastic and glass surfaces for the mobile communications and cosmetics industries. 3206-01

**6 THERMOCHROMIC INK**

Inks that change color or disappear when they are heated above a specific 'activation temperature', then revert when the temperature decreases. They are provided in three standard activation temperatures: 188.2°F/59°C, 190°F/87.8°C and 203°F/95°C. 4422-01

**7 FIRE-RETARDANT INTUMESCENT**

A coating that can prevent buildings from burning. It cannot be ignited but foams up when heated, creating a cocoon-like covering that protects the walls and trim beneath. 4483-01

**8 HYDROGRAPHIC CUSTOM-FINISH PROCESS**

A process involving dipping a base-coated substrate into a water-dissolvable film. The film is suspended on water, with different paint effects produced by the speed and angle of immersion of the part through the floating film. 4607-01

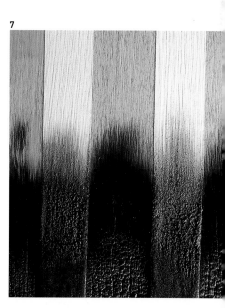

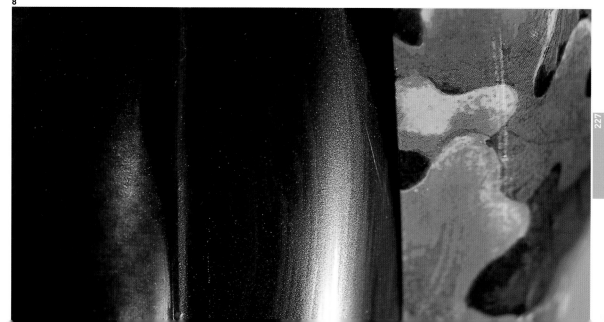

### 1 GRAFFITI-PROOF POWDER COATING

Coatings for indoor and outdoor use composed of acrylic polyester, epoxy, polyurethane triglycidyl isocyanurate or hybrid resin. They may be applied by electrostatic device, hand spray or automatic spray. 671-01

### 2 GLASS VARNISH

A lead- and cadmium-free varnish produces the soft-colored decorative surface effects on this hand-coated glass. Nine colors are available, as well as custom colors. The coating may be applied to flat or curved surfaces. 1505-05

### 3 URUSHI PAINT

Hand-made, uniquely designed decorative objects made by coating materials such as wood, laminated wood, plywood, plastic, cloth, paper, stone, aluminum or copper with a clear, natural lacquer derived from urushi trees. 1835-01

### 4 SOFT-TOUCH PAINT

This polymer-based paint combines a proprietary wax and rubber to create a soft feel. It has been developed for use on wood but can be applied to plastics and metals that have a protective covering. 1509-02

### 5 MICA-PLATELET PIGMENT

Pigments in the form of flakes made of five ultra-thin layers of colorless material applied to a carrier film. The pigments create color by light interference. Precisely controlling the thickness of the layers produces the desired colors. 3077-01

### 6 ORGANIC POLYMER PIGMENT

Brilliant pigments for polymer-packaging applications, developed from organic bases. They are heat- and dimensionally stable even at very low pigment loadings, as well as being FDA-compliant for food-contact use. 2040-09

### 7 SEAMLESS, WATERTIGHT SELF-LEVELING FLOOR

Synthetic, epoxy-based flooring with a smooth, easy-to-maintain finish that can be applied quickly. These anti-static and conductive special-purpose floors are hard-wearing, impact-resistant, hygienic, completely seamless and watertight. 1926-03

### 8 FRAGRANCE-ENCASULATED POLYMER

Fragrance is incorporated into a polyolefin-base polymer and extruded into pellets which form the raw material for secondary injection-molding into various shapes. 3601-01

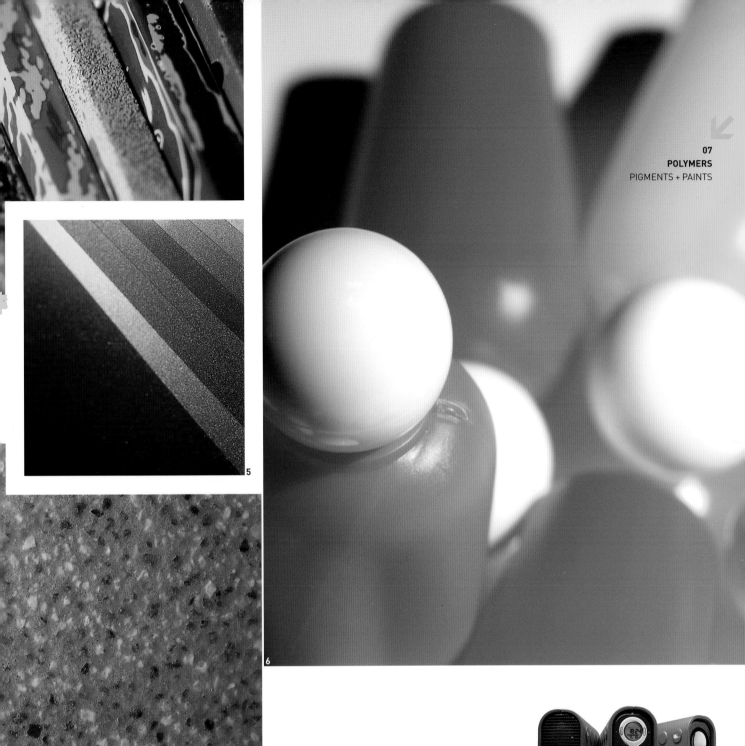

5

6

7

8

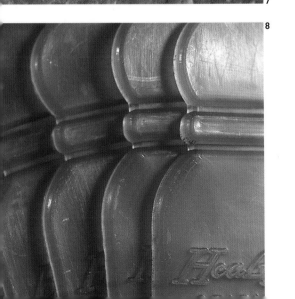

**PHILIPS**
em 2
This durable high-gloss
finish mimics industrial-
molded rubber floor
coverings, here used
on a consumer product.

1 **PIGMENTATION PROCESS**
Compounded-resin technology that allows coloration and special effects to be added directly into pre-molded resin, thereby creating a variety of special effects including custom color, metameric effect, shine, sparkles, tint and edge-glow.   4640-05

2 **HIGHWAY SAFETY-MARKING TAPE**
Highly durable, three-dimensional, reflective and skid-resistant tape that consists of abrasion-resistant microcrystalline ceramic beads bonded in a highly durable polyurethane top coat. Its patterned surface maximizes reflectivity by presenting a near-vertical surface to traffic.  4698-01

2
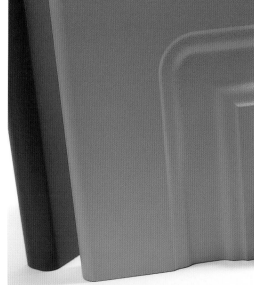

1

4

3

3 **SOFT-TOUCH PAINT**
A highly elastic, leather-like lacquer that is soft to the touch. It is a two-component, transparent or transparent-colored, air-dried lacquer coating with a polyurethane binder. It cannot be recoated or imprinted. 4125-01

4 **POWDER COATING**
A high-quality coating designed for medium-density fiberboard. These coatings can be applied by spray torch. 4839-01

5 **IN-MOLD GRAPHIC PROCESS**
Film-insert molding that incorporates graphics and textiles into an injection-molded part. These polycarbonate and TPU films can be added to create a tremendous variety of graphic and tactile effects during the resin-molding process. 4640-06

6 **POLYCARBONATE EDGE-BRIGHT SHEET**
Fluorescent polycarbonate-based polymers. Polycarbonate resins contain dopants that absorb light from non-visible parts of the UV spectrum, shift it into red, amber, green and violet wavelengths, and emit it from the sheet ends. 4922-02

5

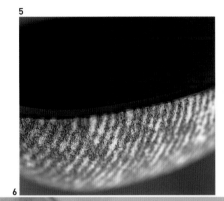

6

231

**1 POST-CONSUMER
RECYCLED WOOD
ALTERNATIVE**
A maintenance-free wood
alternative made from
100 per cent post-
consumer plastic that
can be worked with
conventional woodworking
tools and is resistant to
chemicals, saltwater,
bacteria and insects.
2077-01

**2 *FAUX* WOOD ROOFING
SHINGLE**
A shingle designed to
resemble and replace
wood shake shingles.
It has a Class A fire
rating and the highest
Underwriters Laboratories
impact rating (Class 4), is
freeze/thaw- and UV
radiation-resistant and has
solid color throughout.
4388-01

**3 RECYCLED-RUBBER
ROOFING TILE**
Strong, unbreakable tiles
constructed of steel-
reinforced rubber (84 per
cent) from recycled tires,
with a backing of high-
density polyethylene
(4 per cent) and quarry
slate-aggregate coating
(14 per cent). 4407-01

**4 RECYCLED HDPE SHEET**
A decorative plastic
surfacing material made
of recycled high-density
polyethylene sheets with
a matte finish, available
in various thicknesses and
in patterns, solid colors
and natural. 109-01

**5 POST-CONSUMER
CONSTRUCTION-FORM
SYSTEM**
A lightweight, building-
form system made of
85 per cent recycled
foamed plastics such
as expanded polystyrene,
which is molded into
forms with channels,
compacted and cured.
It provides a permanent
framework for a grid of
reinforced concrete.
4013-01

**6 RECYCLED-POLYESTER
SOLID SURFACING**
Polyester-based material
composed of 50 per cent
recycled plastic. It is fully
recyclable itself. Available
in a standard-size sheet of
2.6 by 9.5 ft/78 by 2.9 m
with a minimum thickness
of 0.47 in./12 mm.
4724-01

**7 ALUMINUM/
POLYSTYRENE
COMPOSITE PANELS**
Composite panels from
recycled bottlecaps, made
without any additives,
which can be worked like
wood and are non-toxic
and water-resistant; for
both indoor and outdoor
furniture. 1907-02

8 **VISIBLE POST-
CONSUMER-WASTE
POLYMER SHEETING**
Scrap and waste materials
such as coffee cups,
CD's, production scrap,
old banknotes and
toothbrushes create the
distinct designs on this
colored sheeting.  4847-01

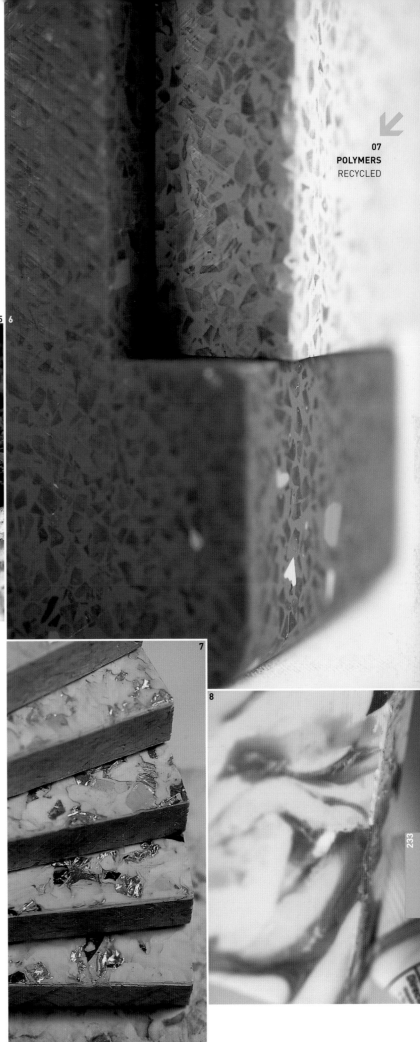

5  6

7

8

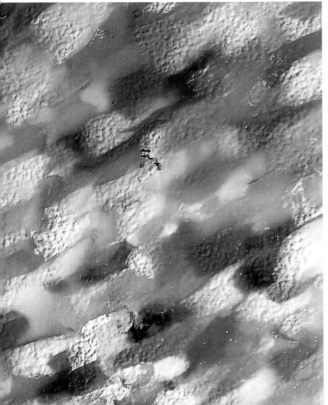

1

2

### 3 SILICONE RUBBER
Compounds for making molds with good tear strength and release properties, which reproduce fine details and are suitable for industrial and art applications including prototype models, furniture, sculpture and architectural elements. 1737-01

### 4 THERMOPLASTIC RUBBER
A recyclable, easily colorable rubber that can be processed on conventional thermoplastic equipment for injection-molding, extrusion or blow-molding. It does not absorb moisture, has a smooth feel, a low impurity and halogen content, and doesn't stain or adhere to silicone rubber. 1764-01

### 1 RESIN SOLID SURFACING
A hand-poured translucent panel embedded with materials inspired by the themes of 'nature' and 'element'. The panel is sandwiched between two clear sheets of polycarbonate or acrylic. 4955-01

### 2 BELTING
Belts made from neoprene, molded urethane, welded urethane (reinforced with Kevlar®), rubber, silicone, Mylar® and other materials. This belting is supplied with or without teeth, reinforcement or elasticity. 4509-01

### 5 HIGH TENSILE-STRENGTH THERMOPLASTIC ELASTOMER
Thermoplastic polyurethane elastomers (TPU's) with high tensile strength; high tear, abrasion, oil, grease, oxygen, ozone, weather and fungus resistance; good damping rebound and elasticity; and low-temperature flexibility. 1849-01

### 6 POLYOLEFIN ELASTOMER
An elastomer based on polyolefin chemistry that, when added to low-cost polymers, improves texture, impact resistance, low-temperature resistance and flexibility. The resin has the malleability of rubber and the easy processibility of a plastic. 2166-01

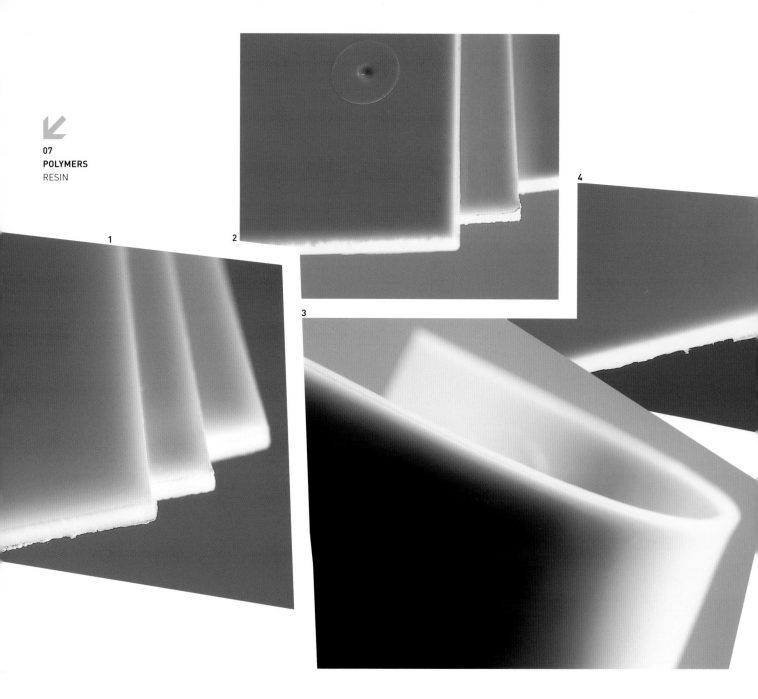

**1 HIGH-STIFFNESS POLYMER**
A thermoplastic resin with a good balance of properties that bridges the gap between metals and ordinary plastics. It offers high strength and rigidity over a broad temperature range, toughness and resistance to repeated impact, and good electrical insulation.
2252-01

**2 POLYAMID THERMOPLASTIC**
Dimensionally stable composite resins made of nylon 6,6 reinforced with a mineral or mineral/glass combination that are strong, stiff, tough and chemical-resistant, and have high heat resistance. The resin is injection- and compression-moldable.
2252-02

**3 POLYESTER ELASTOMERS**
Thermoplastic elastomers that have properties ranging from elastic to stiff depending upon the formulation. These polymers are flexible at low temperatures and have good resistance to heat-aging and to oils at high temperature.
2252-03

**4 ENGINEERING THERMOPLASTIC**
A family of polymer resins for various molding processes. These thermoplastics have a low viscosity suitable for extrusion to produce films, sheet, tubing and monofilament, with properties such as stiffness and toughness variable by composition.
2252-05

5

6

7

**5 POLYURETHANE ELASTOMERS**
Elastomers that are non-halogenated and flame-retardant; burn with low smoke evolution; have good impact and abrasion resistance, low-temperature flexibility, resistance to fuels, oils and greases; and are paintable. For extrusion- and injection-molding. 2939-01

**6 POLYAMID RESIN**
Synthetic resin for high tensile strength and stiff molded parts. This family of resins is based on polyamide (nylon) and includes unreinforced, reinforced (with short and long glass fibers), heat-stabilized, impact-modified and adhesion-modified. 2941-01

**7 VIBRATION-DAMPENING RUBBER**
A vibration-absorbing and shock-isolation material that is a highly deformable solid which retains the memory of the original shape even after repeated deformation and is stable over a broad temperature range. It can be produced in a variety of shapes, sizes and colors, and can be bonded to metals, plastics, foams and cloth. 3030-01

1 **IONOMER FILM**
A family of resins that can be used to produce conventional extrusion/coextrusion-blown film and cast film with good melt strength, toughness, abrasion resistance, clarity, drawability and sealing performance. 3208-01

2 **RECYCLED POLYMER EXTRUSION**
Polymer sections from recycled waste. Waste bags and bottles made out of polyester, polyethylene, polypropylene and polyvinyl chloride are reground and molded into various shapes including planks, rods and bars. 1498-01

3 **RUBBER-MOLDING PROCESS**
A single-mold process that enables variations in color, density, hardness, elasticity and durability to be engineered into specific sections of the product. This material is a form of thermoset elastomeric polymer containing PVC. 3972-01

4 **EDGE-BRIGHT FIBER**
Plastic fibers that emit colored fluorescent light. The fibers (0.01 to 0.2 in./0.25 to 5 mm in diameter) are composed of a polystyrene core with a polymethyl methacrylate (PMMA) cladding that absorbs energy, largely from the non-visible part of the spectrum, and transmits it as visible light from the edges. 4394-01

1        2

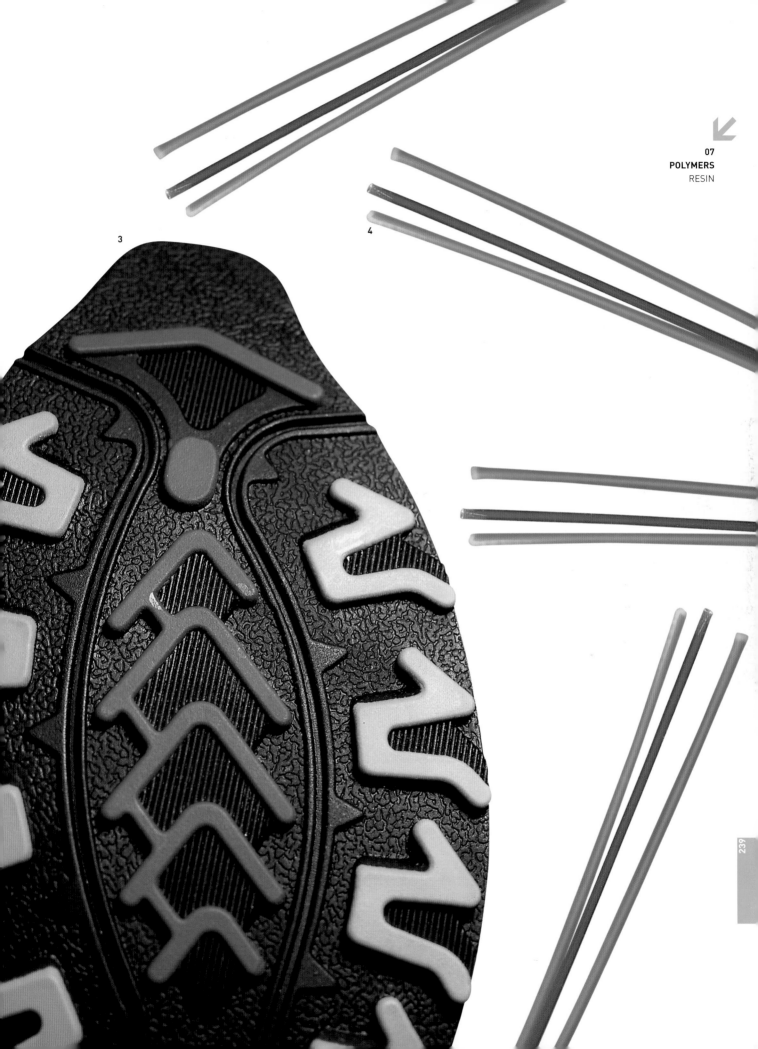

3

4

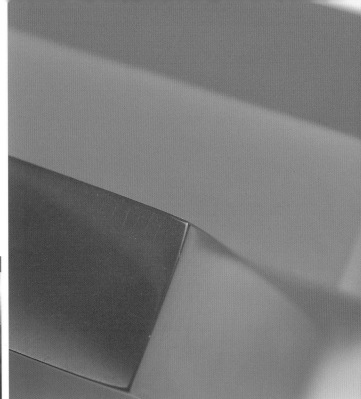

1

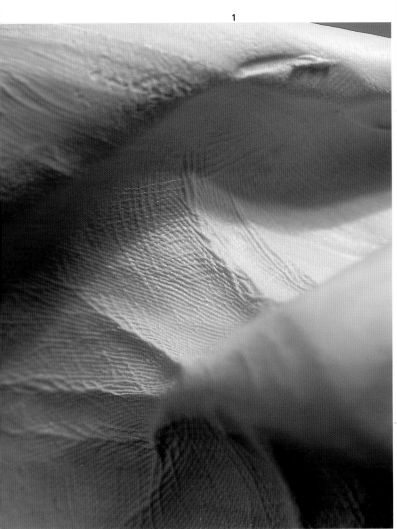

2

3

**1 MOLD-MAKING RESIN**
Fast-drying impression material for mold creation. A highly accurate, smooth impression putty which dries in two minutes or less. This material is able to pick up the subtlest detail when molding teeth, hands and keys. 4616-01

**2 TRANSPARENT RESIN**
Poured resin formulated to yield fully transparent castings. The smooth, resistant, glass-like surface is impenetrable by water and can be used in a variety of indoor applications including tables, sinks, bar tops and wall panels. 4679-01

**3 HIGHLY ELASTIC RUBBER**
Extremely resilient rubber ball made out of 100 per cent Zectron®. Composed of polybutadiene, sulfur and other proprietary substances and originally invented in 1965, it stopped being manufactured in the early 1970s but was reintroduced in 2002. 4624-01

### 4 CONCRETE-MIMICKING RESIN
Poured resin formulated to yield 'concrete-like' castings. The smooth, resistant, rough or pocked surface is impenetrable by water and can be used in a variety of indoor applications including tables, sinks, bar tops and wall panels. 4679-02

### 5 RAPID PROTOTYPING PHOTOPOLYMER
This epoxy-resin polymer that is polymerized and cured by UV radiation is used in rapid prototyping of parts by selective curing, using light followed by rinsing away of the uncured material by a solvent. 4717-01

### 6 HIGH-STIFFNESS POLYMER
Acetal resin that has high stiffness and toughness over a wide temperature range. Available as a homopolymer and as a co-polymer, it has good resistance to repeated impact. It is chemical-, solvent- and fuel-resistant as well. 2252-06

### 7 ACRYLIC SHEET
A cast sheet with a high-gloss surface that is hard, non-porous, impermeable to water and resistant to chemicals, weathering, and infrared and UV radiation. It is used to produce thermoformed parts with thin wall sections. 2950-01

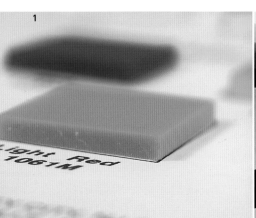

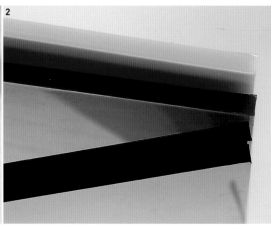

### 1 ACRYLIC SHEET
Sheets of 100 per cent acrylic that are chemical- and impact-resistant and UV-radiation stable. Available in custom colors including brushed and matte metallics, and in patterns and textures. Applications include glazing, lighting and furniture. 3311-01

### 2 THERMOPLASTIC ELASTOMER
An elastomeric and amorphous material. Applications include medical devices, cosmetics, housewares, automobiles and business equipment. 4219-01

### 3 MOROSO
THE BIG E
The rotational-molding process used in the manufacture of this chair allows large, complex hollow shapes to be created. It can only be used with specific resins.

### 4 ATTA
RESIN BAR WITH LIGHTED BULLNOSE
This sinuous bar is cast in resin. The process and the material permit the creation of a seamless, continuous surface limited only by its designer's imagination.

### 5 ATTA
RESIN STAIR
This floating staircase has cast-resin lime-green treads with embedded steel supports. The integral color was created by blending pigment into the liquid-resin precursor before it solidified. As a result, the choices for colors were almost infinite.

3
4 5

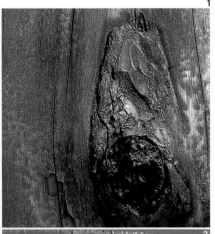

**1**

**1 PHOTOREALISTIC *FAUX* SURFACING**
Molded three-dimensional surfaces that look realistic. These lightweight surfaces are made of glass fiber-reinforced concrete, fiberglass, glass fiber-reinforced gypsum and flexible and rigid urethanes, and include cobblestone, flagstone and slate-roof forms. 4481-01

**2 PRINTED URETHANE TILE**
Patented digital printing of color images onto patented urethane tiling, creating an antiqued 'fresco' effect. 3449-01

**3 HAND-PAINTED REFLECTIVE COATING PROCESS**
Hand-painting a coating of polyurethane or polyester on any custom-designed substrate creates a 3-D effect when light is reflected on the surface. 3901-01

**4 THERMOFORMED NEOPRENE FABRIC**
A breathable, wear-resistant rubber-based fabric composed of microcellular polychloroprene (neoprene), with woven fabric laminates of nylon-polyester/polyester-Lycra® microfiber as lining. 3277-01

**5 FIRE-RETARDANT INSULATION**
A tough, lightweight material made of two layers of aluminum foil laminated to either one or two layers of heavy-duty air-filled polyethylene bubbles. It reflects heat energy and light, and is wind-, water- and fungus-resistant. 3593-01

**6 RUBBER-COATED DOUBLE-WALL FABRIC**
A fabric of polyamide coated with a synthetic rubber. Woven on special tufting looms, it has sixty thousand threads per $ft^2$/two thousand threads per $m^2$. The threads hold the fabric's walls in place at a pressure of 0.5 bar/ 50,000 Pa. 1909-02

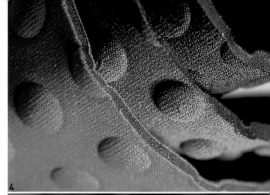

**5**

**6**

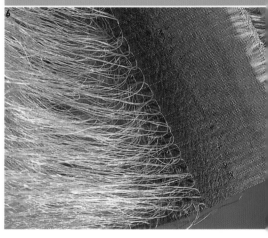

**FIELD TURF**
FIELD TURF
Artificial grass and sod are difficult to make feel genuine underfoot – the cushioning properties of real dirt are hard to replicate. This product uses ground-up recovered athletic shoes to replicate soil, allowing drainage with a sense of spring. The blades of grass are designed to have similar friction properties to natural turf.

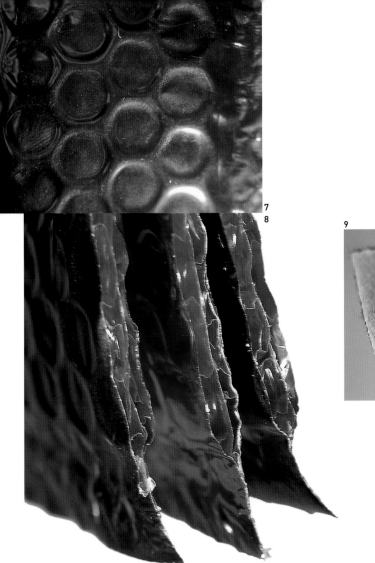

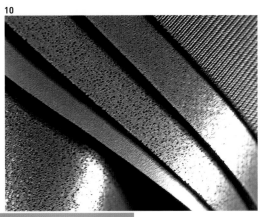

7

8

9

7+8 **ALUMINUM REFLECTIVE INSULATION**
A seven-layer insulation that reflects 97 per cent of radiant heat. The two outer layers of aluminum foil are each bonded to a tough layer of polypropylene. 2679-01

9 **THERMAL-MANAGEMENT APPAREL TEXTILE**
When layered within apparel fabric or integrated directly within synthetic fibers and foams, this encapsulated, microthermal material absorbs, stores and releases body heat. Applications include outdoor apparel, shoes, bedding and protective clothing. 3670-01

10 **THERMAL-MANAGEMENT APPAREL TEXTILE**
Microspheres contained within apparel either absorb or dispel heat depending on the temperature of the wearer, thereby regulating the degree of insulation. 4579-01

**SNOWBIRD**
WATER BOTTLE
Pigments are added to the polycarbonate resin of water bottles to create a low level of fluorescence that gives the water within a slight glow.

1

### 1 ELASTIC UPHOLSTERY TEXTILE

This textile has elastic properties as a result of the weaving technique. Mini-pleated and plain-woven, it is compostable and biodegradable, 76 per cent wool and 24 per cent ramie (a fiber that is one of nature's strongest, with natural stain resistance). 10-03

### 2 WOVEN OPTICAL-FIBER TEXTILE

A durable, flexible, light-emitting panel which does not generate heat or electromagnetic interference (EMI) and which is constructed of multiple (one to six) assembled layers woven from plastic optical fibers with a reflective layer attached on one side. 189-01

### 3 DECORATIVE DRAPERY FABRIC

Sheer fabric woven of slubby, shrinking and shiny linen, polyester, acrylic and acetate yarns. Available in snow, silver and gold. 1975-14

### 4 TUFTED PAINT ROLLER

Pile fabrics with engineered tufted dots or zebra stripes. Composed of 100 per cent polyamide with a 100 per cent polyester backing that may be thermofused, glued or sewn, they are manufactured for use as paint rollers to produce patterned paint effects. 2394-03

### 5 DECORATIVE DRAPERY FABRIC

Innovative woven and knit fabrics of polyester, nylon, viscose and silk that are sheer, pleated, crinkled, crushed, taffeta, organza, laminated, bonded, coated, printed, flocked, seersucker, warp-printed, iridescent and metallic. 3446-01

### 6 TEAR-RESISTANT NYLON

This semitransparent fabric has eye-shaped accents created by printing an elliptical shape in a pastel color and then removing the bottom half by laser cutting. 4438-01

### 7 SELF-SUPPORTING WOVEN TEXTILE

Tough and durable membrane textiles for conventional upholstery for chairs and other seating without supporting elements, these contain some elasticity and have high tension and tear resistance. 3580-02

### 8 WOVEN FIBER-OPTIC TAPE

0.03 in./0.75 mm-diameter fiber-optic threads woven into nylon webbing that emits colored or white light when connected to a battery or mains power. 3829-01

### 9 DECORATIVE DRAPERY FABRIC

An innovative fabric made of cotton and polyurethane. Weft-inserted polyurethane-coated wires create texture in the woven fabric. 4438-02

### 10 THERMOFORMED WOVEN-POLYMER NET

A 100 per cent polymer net with three-dimensionally formed shape. The regular, repeating nodules within the net are heat-formed during post-processing. 44-03

### 11 WEAR-RESISTANT UPHOLSTERY TEXTILE

A very durable, knitted, mesh-faced, layered textile which has a 100 per cent polyester-mesh face laminated onto a 100 per cent polyester tricot back and Teflon® finish/backing. 314-01

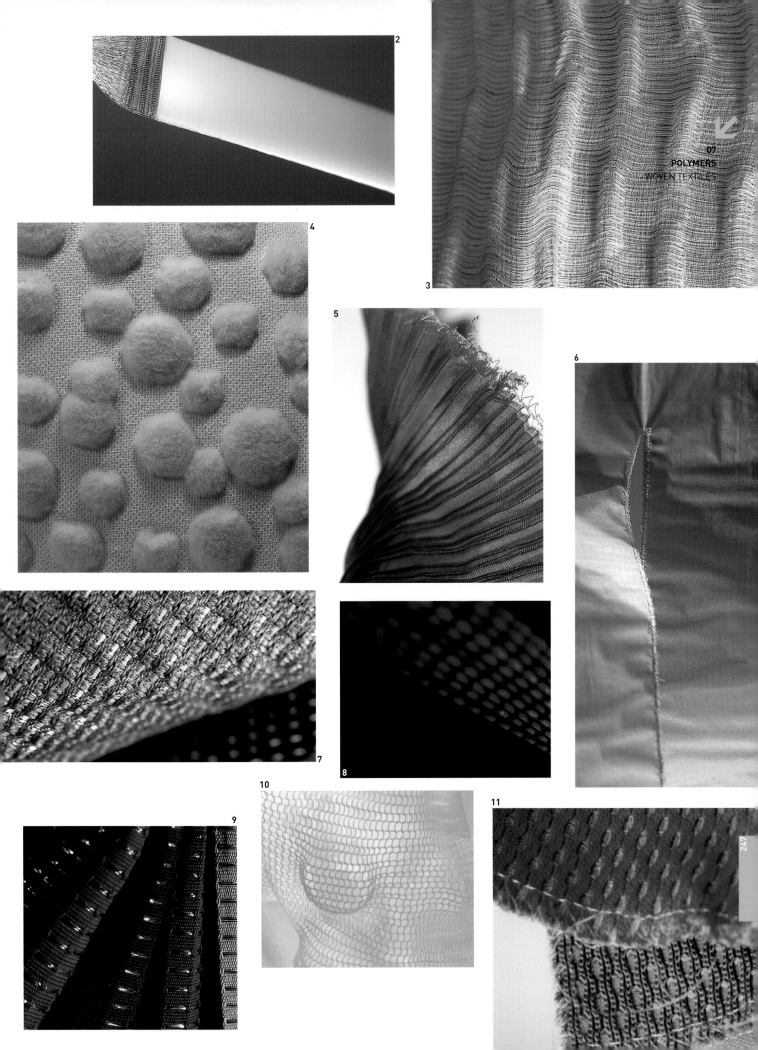

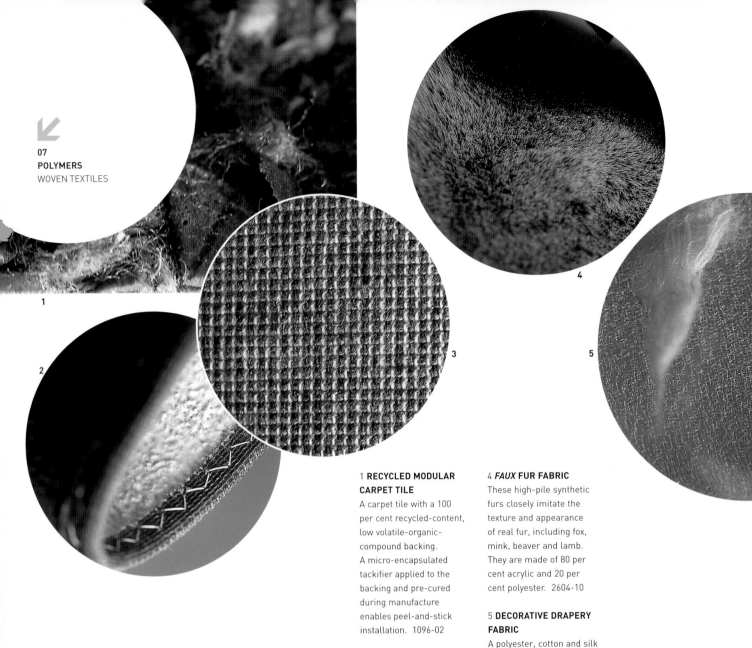

1

4

3

5

2

**1 RECYCLED MODULAR CARPET TILE**
A carpet tile with a 100 per cent recycled-content, low volatile-organic-compound backing. A micro-encapsulated tackifier applied to the backing and pre-cured during manufacture enables peel-and-stick installation. 1096-02

**2 SOUND-ABSORBING WOVEN CARPET TILE**
Non-flammable, sound-muffling wall carpet that feels like wool and has a flatweave texture. Woven from a blend of 80 per cent wool and 20 per cent non-static nylon. 1104-01

**3 RECYCLED SODA-BOTTLE DECORATIVE TEXTILE**
Soda bottles are the source of the 100 per cent recycled polyester used to fabricate these nine decorative textiles. Five are designed for panels and four for drapery. 1940-04

**4 *FAUX* FUR FABRIC**
These high-pile synthetic furs closely imitate the texture and appearance of real fur, including fox, mink, beaver and lamb. They are made of 80 per cent acrylic and 20 per cent polyester. 2604-10

**5 DECORATIVE DRAPERY FABRIC**
A polyester, cotton and silk decorative fabric. Cotton and silk slubs are woven directly into a polyester fabric to create soft, thick nubs that protrude at random from the surface. Available in 54 in./ 137.2 cm-wide bolts. 1975-22

**6 ANTI-MICROBIAL FIBER**
Fiber formed by the extrusion of an acrylic polymer with Triclosan anti-microbial and Tolnaftate anti-fungal additives. The addition of the additives in the polymer's molten state imparts inherent antimicrobial properties. 4318-01

6

**DESIGNTEX**
IN SPACE
Evenly spaced polyester-
cushioned ribs project
from the base of this
three-dimensional spacer
fabric. This technology,
employed extensively in
sneaker manufacture,
is used here as knit
upholstery.

1

2

3

4

**1 NON-PVC CARPET BACKING**
A thermoplastic olefin that does not contain PVC composes this carpet backing. After it is applied in a molten state directly to the back of the tufted fabric and flash-cooled, the carpet can be processed into tiles or rolls.  3296-01

**2 TAFFETA DRAPERY FABRIC**
The luster of this heavy acetate taffeta contrasts spectacularly with the metallic polyester threads on its surface.  3495-02

**3 TEDDY-BEAR FABRIC**
Decorative textile for upholstery applications. Attached to the cotton backing are 100 per cent mohair fibers, the tips of which have been burned and melded with acid to lighten their color. The textile is available in 98.2 ft/30 m lengths and 4.6 ft/1.4 m widths. 4689-01

**4 SUSTAINABLE FABRIC-PRODUCTION PROCESS**
This process for producing fabrics (available for licensing) includes established sustainable methods – such as eliminating waste, making emissions less toxic and reducing energy demands – to create woven textiles from three types of recycled fibers.  4128-01

**5 THREE-DIMENSIONAL TEXTILE-PRODUCTION PROCESS**
A recently developed knitting machine that can produce three-dimensional textiles with novel properties such as preformed shapes with a wide range of different profiles. Applications include stitch-bonded fabric components for engineering, automotive, aerospace and sports, and foamed or coated and inflatable shaped parts. 3824-01

**6 RECYCLED PANEL FABRIC**
Industrialized fabric for panels, upholstery and upholstered walls. This textile is composed of 51 per cent polypropylene tape and 49 per cent recycled polyester with an acrylic backing. Available in 66 in./1.7 m widths, it comes in eight standard colors.  1940-06

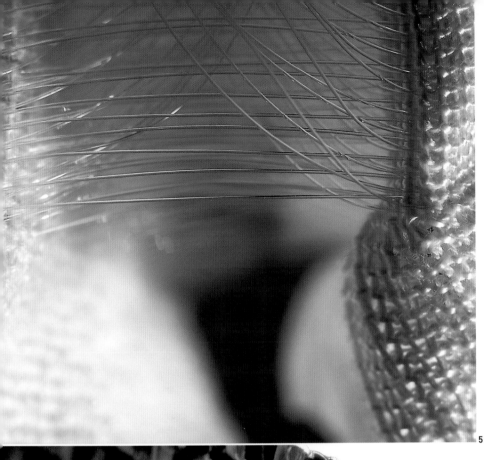

5

6

# WHAT THE MATERIALS GURUS SAY
## A QUESTIONNAIRE

IN THE COURSE OF PREPARING THIS BOOK,
MATERIAL CONNEXION ASKED FIFTY-FOUR CUTTING-
EDGE DESIGNERS AND ARCHITECTS TO PAUSE
TO CONSIDER THE ROLE OF MATERIALS IN THEIR
WORK – TO THINK ABOUT WHY THEY LIKED TO USE
PARTICULAR MATERIALS, WERE INSPIRED BY THEM,
PREFERRED THEM TO OTHER ALTERNATIVES, AND
FELT THEY WERE SIGNIFICANT AS REFLECTING
SPECIFIC VISUAL, TEXTURAL, FUNCTIONAL, CULTURAL
AND OTHER VALUES. EACH DESIGNER WAS ASKED
THE FOLLOWING QUESTIONS:

**1**

**TELL US ANYTHING YOU WANT ABOUT HOW MATERIALS INSPIRE YOU,
ANY IDEAS, WISHES, PROVOCATIVE STATEMENTS IN REGARD TO MATERIALS.**

**2**

**WHAT IS THE ROLE OF INNOVATION IN MATERIALS OR MANUFACTURING PROCESSES IN YOUR WORK?**

**3**

**WHAT IS YOUR FAVORITE MATERIAL? WHERE DO YOU SEE THE MOST POTENTIAL?**

**4**

**WHAT WOULD BE YOUR DREAM MATERIAL? WHAT PROPERTIES WOULD IT HAVE?**

THEIR ANSWERS FOLLOW.

## PETER ARNELL

**2** Curiosity. Surprise. Delight. Desire. For me, the ultimate role of any material or process is to enable manifestation of new experimental value. We all develop elemental understanding and expectations of material throughout a lifetime of interactions. It's when we challenge those notions, expanding our thinking and engagement with an idea or object, that it's most rewarding.

**3** I tend to use materials that enhance the expression of an idea. It's fun to watch a thought magically transformed through form and material. Favorite? I don't have one, but currently I'm into utilizing light. So I'm fascinated with what may come after 'smart' materials. As materials, sensing, computing and interface converge, I can imagine interactive materials. A product might someday adapt to my touch. Interfaces may provide unique sensory responses to my personal context. 'In the eyes of the beholder' could take on a whole new level of meaning.

**4** I've rarely had a material let me down in application. I suppose I dream for better material management from cradle to grave to cradle etc. Certainly materials can assume certain roles and responsibilities in reuse and recycling, but the difficult issues are in human and societal behaviors. In *Earth*, David Brinn describes scenarios where we are eventually mining landfills for materials. Very plausible, but how can we avoid going there?

## GIJS BAKKER

I have a strange attitude toward materials. On one hand I am fascinated by them, on the other hand I am afraid of their beauty. Maybe it's because of my background. Trained as a jewelry designer, you learned that gold and precious stones are beautiful. I recognized this beauty and it's for this reason that I have always hesitated to use them, only when it's absolutely necessary to express my thoughts. Materials have never led me to a new design. Already from the beginning of my career I experienced the concept (thought) as the only thing that matters. I used to say the form (material) is only the wrapping of the idea. A concept is like a dream. Not connected to the heavy reality but going beyond the reality in absolute freedom. This gives me strength and power. Of course then it all depends on your skills and knowledge about materials and techniques, to be able to materialize your concept in the most adequate way.

When doing projects with Droog Design, sometimes the materials are the starting point, such as glass, ceramic or high-tech fibers. The first thing we do during the brainstorming sessions with a group of young designers is to get rid of the sentiments connected to those materials. We start to philosophize about our relation in terms of sensitivity toward the given materials. For instance, when we were commissioned by Rosenthal (a German porcelain company), the first thing that appeared to us was our dislike of the over-perfection of their products, no scratches, no cracks in the glaze, only perfect forms, but for us without life. This gave us distance and freedom to experiment and to come up with completely other ideas, such as random decoration, glaze looking like knitted cotton, and to introduce forms from other disciplines in the world of porcelain.

## CHRIS BANGLE

**1** I am fascinated by the use of the material properties per se as the design statement, instead of properties being a 'way to adapt' a material to a given function. And if 'glass and light are two forms of the same thing' (Frank Lloyd Wright), what are the *doppelgänger* forms for metals, plastics etc.? Perhaps if we understand these 'inner spirits' (glass – light etc.) we would better understand how to approach their application and integration into design.
**2** Very important! We depend upon an authentic and effective connection between form and content, and if anything the integration of materials innovation into the car world has become the bottleneck we must improve … it is too slow in happening.
**3** Materials with multiple functions and multiple 'lives' are most interesting, yet understanding an 'old' material in a new way can be just as inspiring. I think we must approach it in both ways: what is new and offers something and what is known that requires thinking outside the dogmas to discover new potentials! What are really stimulating are materials that change our dogmas concerning 'meaning'. Some time ago, we had a project to substitute real wood in cars with something less expensive but just as authentic, and with a similar 'meaning' to our customers. The material we were searching for we simply called 'home', for

all the associations wood brings us. I'm still looking.
**4** On a practical side, whatever is of low investment on the process side, but with high performance and new function and emotionalism on the product side, is very interesting.

## CINI BOERI

**1** I love materials lightened (in the sense of being less heavy) like stone, wood, metal, when they are not full of bulk and heft and mass. I love the idea of using them in an easy way so that they become almost weightless.

I also like chromium-plating, but it's very expensive. I'm waiting for the not-expensive version.
**3** Glass and everything transparent.
**4** I would like to be able to turn glass from transparent to opaque.

## LAURA BOHN

**1** When we begin a new project, 90 per cent of the time we start with a material we've fallen in love with. It might be something we've used before, or something we've never tried. It's not always successful, and when it's not, it's usually as a result of the physical properties of the material. We don't always know how a material will behave in a given situation. Because I'm a designer I'm always trying out new materials on myself, in my own spaces.
**2** We try to use them all! We see the role of the designer as figuring out what something is suitable for. Lots of these materials don't yet have a use. What are

the conceivable forms? What makes this material useful?
**3** I am drawn to man-made materials. I try to use them over stone, marble or granite. I love recycled materials, epoxy and composites. My favorites are man-made but green. Unfortunately a lot of these new products are still beyond the price point where it is feasible to use them in large quantities.
**4** Something impervious that wouldn't scratch or stain, resists heat, and is inexpensive and beautiful. Durable. Fake.

## MICHAEL BRAUNGART

**1** I am a material boy. The path to a healthy, abundant and sustaining world is not dematerialization of our industry and our products but Re-materialization. When materials are healthy, well performing and environmentally beneficial, they become nutrients, and they become part of an industrial metabolism that renews and regenerates people, the environment and the economy.
**2** Supporting and applying innovations in materials is 90 per cent of our work. New materials are key for businesses, because they create a key source of competitive advantage over low-quality, cheap imports from Asia.
**3** Magnesium and zinc are my favorite substances, because they are truly nutrients, and there is a deficiency in the uptake of these minerals within the United States and Europe. Polymers are also an exciting area. In this context, polysulfonic compounds and polylactic acid (when it is free from GMO's and does not compete with food sources for raw materials) are ideal materials and ideal

nutrients for industrial and biological systems.

**4** An ideal material would be one whose every ingredient is positively defined to be beneficial to humans and the environment. This could be either a biological nutrient, which can become compost after use or safely be used as a fuel, or a technical nutrient, whose properties enable the material to perpetually flow in industrial cycles.

## ANDREA CANCELLATO

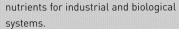

**1** People see materials all around, but they seldom look at them closely. It's incredible what new universes you can discover by asking yourself, 'What is this made of?' If you have someone who can answer the question, it's difficult to stop the human mind from trying to imagine how and where else you can use this material: in a flash you become a designer!

Usually we say that materials are inanimate matter, and yet they provoke such a wide range of feelings. Perhaps we have become accustomed to thinking about them the wrong way.

**2** Materials are able to provide many tools that I use daily in my work, from the invisible copper wire crossing the walls in order to supply energy to my computer to the wood of the chair I sit on. Our world is material.

**3** Ceramics are my favorite materials, as they represent the oldest matter that man started to work, starting from 'dust' and water. I like the potential to shape by hand something that when finally fired becomes hard; it is very stimulating. It's a pity that in today's life we've lost the time

for experiences such as this one which combine creativity and relaxation at the same time.

**4** I have never liked the way that walls create such a permanent and irreversible barrier. I would like to have concrete that can become as clear as glass so that I can open all my rooms to the warmth of the sunlight during the day, and then when desired it can become opaque again, protecting my privacy at night, letting me feel more safe while I sleep.

## CLINO TRINI CASTELLI

**1** I always consider materials as the mother of all form. If we design only with light, sound, texture etc., that necessarily makes the material the protagonist in the story. I have been working in that way since the early '70s.

**2** The manufacturing process is fundamental. I always consider the way in which materials are manufactured to be one of the most thrilling design generators. Think about synthetic fibers as an example of innovation, one that has had a profound impact on the larger design community.

**3** The last material with which I worked is always my favorite. It's a very parental approach, and an active process. I try to be objective with color, but there are periods when I see particular color languages, like musical chords, that are more appropriate for that moment. But like a mother, I tend to consider even the difficult children something to take care of.

In the '90s the greatest potential was in the rediscovery of synthetics. Then there were natural materials, like for example

linoleum, that appear to be artificial because of the way in which they've been treated. Today I'm most interested in hybrids – veneers of different materials and composites. I'm intrigued by both the processing of the material (in the sense of cutting it, supporting it, treating it) and the properties that come from its use in combination with something else, such as greater structural strength.

**4** Some sort of anti-matter, slow, colorless anti-matter, something that absorbs, digests and dematerializes everything with which it comes into contact. Slow enough so that you can put your finger in and pull it back out.

## GIULIO CASTELLI

**1** Materials are the essence of every single project I have ever worked on.

**2** Responding to this question is like opening a door that is already always open. In 1949 I founded Kartell with the philosophy (that still endures) to only use new materials and new technology in the designs.

**3** My favorite materials are the plastics, the polymers. They are truly the materials of the twentieth century – and of the twenty-first: there are so many yet to be discovered or developed.

**4** A heat-resistant plastic, one that won't melt, burn or catch fire.

## ERIC CHAN

**1** Less is More. Dematerialization is the trend of the future. Man consumes too much material and expects too little performance from it. We will begin to expect greater performance from products by using the minimum amount of materials possible.

**2** Materials are essential to my design process; we always push the manufacturing and technology envelope to give us an edge. Consumers expect new sensations, but most of the mass-production process is already standardized. New materials are key for future design innovation.

For example, we are currently working on a kitchen product that requires a new material to give us shape-morphing characteristics for handle comfort and color changes for heat safety – all while being dishwasher-safe and bacteria-resistant.

**3** Aerogel. Invented by NASA, it is extremely lightweight (lighter than air), strong, an excellent insulator and has numerous applications – from a super-light ski jacket weighing less than an ounce, to incredibly compact/thin wall coolers, to a smaller alternative to silicon in computer chips.

**4** The best material is no material. Objects lie idle for most of the time, so we should expect more from material efficiency. I would love to see a material that exists only while the product is performing its function, then vanish into virgin form after use.

## PATRICIA CONWAY

*Materiality* The quality of being physical, real, perceivable by touch; the quality of having substance ...

**1** Do these qualities inspire design ideas the same way in the context of practice as they do in the context of teaching?

Probably not. For most practitioners the choice of materials is dictated by the specifics of a form or object being developed in response to a client's needs or desires; then ratified (or not) by that client's budget and performance requirements. In a teaching studio, both students and instructors are encouraged to exercise their own imaginations, to test rather than automatically accept application limitations, to risk failure rather than guarantee success. This is why so many architects and designers want to teach despite the resultant pressures on their practices. The teaching studio, much like the laboratory, is the place from which truly new ideas are most likely to emerge.

**2** When it comes to innovating materials or advancing manufacturing processes, how do the roles of the thoughtful studio instructor, together with a group of talented students, compare with the ambitious practitioner supported by his/her staff? Here it seems that the practitioner has a decided edge. Innovation and advance are driven by a real need to solve real problems in real time. With the possible exception of well-funded school-based design-research programs, it is only viable design offices that enjoy the buying power and market access essential to stimulating innovation and advance on the supply side.

**3** My favorite material is water. While clearly to be avoided in basements, water has more design utility and aesthetic range than any other substance on earth. It generates power; it heats, cools and moves. It can be structured through pressure or dissolved into mist. It can be smooth, rippled, undulating, cascading, colored, flavored, infused with scent. It

can be hot, cold, noisy, quiet. It can splash musically or be 'tuned' to break up unwanted sound. It can float tiny candles on a dinner table, wash a large expanse of wall, steam in a tub. Name one other material that can do all that (and I could go on).

**4** My dream material would be something that never gets dirty and glows in the dark. I would use it to line pocketbooks, purses and bedroom slippers.

## ORLANDO DIAZ AZCUY

**1** From the very conception of the design, the materials inspire me. I do not wait to select a material after I have an idea, they come out together. Materials and design are the 'chicken and the egg' of my expression.

**2** I always look to create something new, but many times it is not in the possibilities of the project. New materials are always a good beginning to achieve some creativity.

**3** Indoor wall materials. When concrete is poured, the end result is there, with no more work necessary. Unfortunately there is no alternative to gypsum board. The only possibility is to decorate it with fabric, wallpaper, paint etc. There should be more wall materials that provide unique finishes at the same time.

**4** An acoustical sound-proofing material. When spray ceilings were discovered, it was a very successful material but ugly. We don't have a material today that has the same acoustical properties but with a high aesthetic appeal. The closest is stretched fabric such as 'Eurospan', but even this has a lot of limitations.

## TOM DIXON

**1** All designers work from the same departure point, an obsession with the properties of the material worked with: the ductility and tensile strength of steel; the hardness and clarity of glass; the flexible yet complex nature of wood; or the way polymers can be both versatile and plastic. The exploration of these characteristics forms the inspiration for all my products.

**3** It would be madness to restrict oneself to a single material. I worked a lot initially in metal, for its versatility and strength. I work a lot in plastic now as it is the authentic twenty first-century material.

**4** My dream material would be a polymer that was made from renewable sources which would biodegrade to order.

## ROBERT EBENDORF

**1** I am creatively inspired. Visually by the materials that are abundantly around me. I oftentimes am drawn to colors, textures and materials that I find out and about in our universe. For example: objects that have been discarded in parking lots, alleyways or construction sites. With closer viewing I oftentimes will gather and bring these materials to my studio, and the visual creative process has begun.

**2** As a goldsmith, the manufacturing processes that I use are my hands, measurements and the techniques that a goldsmith traditionally would use. It is craftsmanship and paying attention to details and the playfulness between innovation and creative thinking.

**3** I have no favorite material. I find many materials to be extremely exciting to work with. A crab claw from the beach, broken car-window glass picked up from the street or perhaps an animal carcass that I find in the woods. Beauty is in the eyes of the beholder.

**4** The material must intrigue me perhaps by color, hardness, flexibility. These properties set my creative juices running and the hand investigates the materials, and the designing process begins. There are so many possibilities available in the forms of found objects and traditional materials that I would not want to limit my dream materials to just one.

## ANNA CASTELLI FERRIERI

Almost all my design work and my typological inventions were inspired by the avenues I saw opened by new materials. New industrialized production technologies created a fundamental leap in quality – better performances and new potential all devoted toward a richer design. This happened for me many years ago, when plastic materials started coming at a rapid pace – a new type of plastic with different inherent possibilities every year – and industrial design began its important story.

Now I think this concern for new materials and technologies is even more fundamental, owing to the spectacular changes in our way of life, which I must say is not always positive, but which we have to think of if we want to be alive in our world.

I believe we will soon be seeing our dream material, which should be so light and strong as to bring us easily to all the planets of our Milky Way.

## RENÉE FORD

My interest in materials reflects my practical nature. Creating and producing a new material should not waste precious non-renewable resources; the material should be environmentally benign to make, to use and to dispose of; and it should be economic to manufacture and use. Cradle-to-grave thinking is essential when developing a new material. And most important to me is that a material should serve a useful purpose.

In my present work we are creating and developing a new construction material that is fire-resistant. Its original use was for molds, and artists like its durability for sculpture. Our innovative version of this material, a fire-blocking composite which is in the Material ConneXion Library, embodies all of the above concepts and an additional significant one – it will save lives and property.

Many years ago, when plastics were just beginning to become acceptable as useful materials, I wrote, 'One ideal plastic – strong as steel, clear as glass and cheap as dirt – is still the stuff that dreams are made on.' In the real world we work with many hundreds of materials, each with distinctive properties, all falling short of some ideal. This motivates innovation and brings new and exciting materials to the market.

## EMANUELA FRATTINI

**1** When I see a material that is interesting on an object, my first thought is of translating its properties into totally different applications.

**2** Innovation can consist also of a novel

application for a traditional material, but in general material research and knowledge are vital.

**3** Still leather and wood, because they allow low-tech solutions (low investment) – large potential; potential also in all alloys or composites that combine desired properties.

**4** A material that is inexpensive, recyclable, has a broad range of applications and replicates the pleasantness (tactile and smell) of cashmere or leather.

## NICHOLAS GOLDSMITH

**1** How do you use the least amount of material possible for your design? At a certain point, this literal minimalism creates a tenuous relationship between the different materials. One ounce less, one cable less or one drop less, and the whole structure falls apart. It is the moment when every material is working at its hardest that I try to preserve, because at that moment, a harmony exists between all the parts – a natural equilibrium which embodies a remarkable beauty.

Circus structures have always fascinated me, where everything is treated as a precious commodity: the show elephants raise the big top, the poles also support the trapeze rig and its netting, the lighting rig is used to raise the center fabric etc. It is a blueprint to look at for the future, where the cost of materials will be at a premium, and the least quantity makes the most sense. As a result, materials such as woven skins have always inspired me. They take their strength not by piling up more and more material until it becomes a block but by applying tension in the surface to create a stable structure. Taking such a material and spanning two hundred feet illustrates the phenomenon of how the lighter an object is in reference to its load-bearing capacity, the more efficient a material becomes.

**2** Each material has an inherent strength and weakness and can be joined to other materials in very specific ways (detail connections). For us, for example, when we were asked by NASA to develop a deployable air-lock for the next-generation Space Shuttle, we were intrigued by the challenge of meteorite-proof fabrics and whether some of the high-performance fibers such as Kevlar®, or Vectran®, could be used for this application. Material testing in various stages of development allowed us to feel comfortable in using a medley of layers of different fabrics. This process was both fun and educational for our office and created a cross-fertilization of this technology on other completely different applications. Maybe the large braiding system we ended up using can work for small fabric office spaces or some other use where it has never been applied before.

**3** This month I am fascinated by expanded PTFE (Teflon®) wovens. They have the hand of a silk yet the durability of the longest-lasting fabrics (thirty+ years). They have translucencies of up to 40 per cent yet strength sufficient to cover a stadium. However, in the long term, my interest is in smart fabrics. Weaving allows each fiber to be different; it allows for information to travel along each length and change color, form, stiffness etc. There are fabrics that can change colors, fabrics that can become stiff like shells under human contact, fiber-optic screens that can be manipulated; it all leads to spaces surrounding us that are made of fabrics which are multi-function, multi-layered and multi-use.

**4** My dream material would be a long-lasting structural fabric with adaptable translucencies from opaque to clear depending on the solar condition, that would be 100 per cent sustainable (recyclable with no toxic off-gasses from the chemical manufacturing process), and that would be able to act as a photovoltaic surface to generate its own power. We have materials that can do each of these things separately but none that can do more than one successfully.

## CHARLES GWATHMEY

**1** I am inspired by materials that effect a transformation – that endure but can also patinate over time and maintain their perceptual as well as their material integrity.

**2** We must become more research-oriented to fulfil an idea/intention into reality. Historically we have invented and experimented specifically.

**3** Zinc, resin, patterned/structured glass. Transformative readings: mesh – composite.

**4** Depth of reading: solid, translucent, layered, with high strength and structural integrity. Transformation of natural materials: aluminum, leather etc.

## ZAHA HADID

**1** The choice for a certain materiality of an architectural object follows the initial architectural concept and the resulting formal approach that is applied specifically to each project.

In a design environment dominated by new softwares that enable us to rethink form and space radically, there is an urgent need for materials that match our computer-generated complex shapes and spatial conditions. After simulating demanding geometries on the computer's screen, the search for materials with matching properties starts. Material properties might be explored in this research process that had not been thought of. At best there might be a new application or construction method for a material as a consequence of our research. We aim for an expansion of the material's performance and try not to think in the limits that are given to a certain material by conventional applications.

**2** Starting from small-scale representational or working models, rapid prototyping methods are more and more becoming part of our design process. CNC milling, stereolithography, laser sintering and 3-D printing are very useful innovations that enable us to turn a computer-generated design into a materialized reality. The precision that can be achieved via rapid prototyping helps to control the design from the very beginning.

At the same time, full-scale models or built environments might be constructed with support from the very same rapid prototyping methods, so there is a congruency between the small-scale models and the actual building.

Techniques that enable us to 'stretch' a material – like thermoforming – in order to achieve highly complex double-curved surfaces are currently being explored for domestic interiors as well as for façade claddings. On the one hand, we can enhance ergonomic qualities for interiors; on the other hand, these techniques allow us to produce a completely new formal expression in the case of external applications. Buildings might appear like huge masses of melting ice, e.g., and thus add a strong poetic quality to the urban space. New manufacturing processes play an important role in the creation of 'liquid spaces'.

**3** Potentially materials with multiple properties will become more interesting in the future. For shell structures we are testing fiber composites at the moment. A monocoque shell made out of a fiber composite is structurally performing very well, although its weight is minimized. At the same time these structures have very good insulation properties and can take any color or surface finish you like.

Another material that leads us to a futuristic level of discussion is fiber concrete. This material can possibly take any shape depending on the casting methods that are applied. At the same time it has a very good structural performance since it can withstand tensile stress. Strength and formability are combined into a single material layer.

There has been ongoing research with thermoformable materials for some time now in the office. Tested on furniture-scale products, solid-surface materials are used to create domestic objects with a high ergonomic quality and an extremely smooth appearance. Lately we created a domestic environment by use of thermoformed solid-surface materials that unfolds all functional necessities out of a single seamless surface.

**4** In regard to the discussion of a sophisticated architectural skin, this would be a material that can be twisted, stretched, bent and wrapped around in whichever imaginable way and which at the same time can be transparent or opaque, structurally self-supporting, and take any surface quality or color one can think of. This material would have multiple properties all compressed in one single layer.

## PAUL HAIGH

**1** I look primarily at the intrinsic qualities a material may have – e.g. the subtleties and depth of a particular stone or the uniqueness of a man-made material to evoke new thoughts on pattern and texture. Potential for manipulation of application and fabrication are also important in giving the material a context for exploration.

**2** The architectural culture is primarily one of material 'specification' in which innovation is done by the 'product' manufacturers; our material invention in architecture comes largely though 'material' manipulation or unusual applications.

In product design, I see greater opportunities for material invention by exploration of the processes used in either the fabrication of the product or the material. I have been more curious about creating 'mass-produced' products that through innovation in materials application or processes retain a sense of uniqueness rather than sameness.

**3** Currently I would say that glass is high on my favorites list. My experience in producing designs in this 'oldest of materials' has led me toward a renewed sense of optimism about the transformation of light in space or product.

**4** This question raises the notion of some sort of perfect material. I don't know what a perfect material would be. I enjoy the margin between perfection and imperfection.

### SHEILA HICKS

**1** When a material expresses its natural behavior, I observe and honor it. Then, as an afterthought, I contradict it and discover its alter ego.

**2** They are capital! Nothing materializes or 'gels' until I meld an idea with its body: *faire apparaître*, *se matérialiser*, *prendre forme*, *prendre corps*. Materials and fabrication processes must continually stretch to catch up with our rampant imaginations.

**3** Linen, cork, metal fibers, stone strands.

**4** At the moment I am following the process of creating fiber from rock. Then I will weave a steel. Its properties: water, fire, wind and fashion resistance.

### KARA JOHNSON

Materials are an intrinsic element of any object or space that is physical in its nature. But materials are more than the science or accident that develops each one – nylon, titanium, alumina, silicone.

Materials are about color – the color of aluminum or stainless steel; the natural color of nylon; the subtleties between white polyethylene, white polycarbonate and white elastomers. Materials are about surfaces – anodized aluminum that has been polished by hand or lightly etched; textures that are created in the mold or by pressing fabric onto a soft surface. Materials are about structures – layers of materials that are constructed from sheet- or injection-molding; simple forms that are molded from felt; frames that support or suspend fabric. It is these elements that inspire designers to create products that connect people with objects or spaces.

Materials and manufacturing technologies are seamlessly integrated into the process of design and innovation at IDEO. These technologies support us from first concept to final implementation. Another point of innovation for designers and materials today is in the process of the design of materials themselves. In close working relationships with material suppliers, IDEO is experimenting with technologies that will define the material opportunities that are offered to designers in the next years. Current trends in design will inspire the creation of materials that have high contrast or are invisible; integrate technology or are natural, not synthetic; age with dignity or are defined by the person who touches them.

### WOLFGANG JOOP

**1** It's like a cross-fertilization between my vision and the material. Sometimes I'm looking for something I can use for my creations. Other times I find a material and alter my creations.

Any kind of material has a soul. You just have to listen to what the soul is telling you, to its secrets and desires.

To me it's a mystery – a fabric tells me how it wants to be molded like a fish tells me it's time for it to get out of the pan when I'm cooking.

**2** If you see a new material, it's like a key that opens a door. Behind the door there's a kind of new world, filled with new people, new figures, new ideas, new visions.

**3** I'm often very surprised by the smartness or intelligence of new fabrics. Right now, I'm very inspired by paper tweed for summer, because paper and couture are a rather controversial combination. The fabric is light, cheap, humbling but worked in an expensive way. You can add things to make it very luxurious.

**4** My dream material would be water-, stain- and liquid-repellent. Warming and cooling at the same time. And edible after it's worn out – or out of fashion. Whichever comes first.

### GERE KAVANAUGH

**1** If you don't have materials, what can you do? My relationship to materials is like sense memory – I have a very collective mind. When I see something, I don't forget it. I'm able to go into my own mental library and put materials together quickly, with their histories.

**2** The new use of materials is an obligation for designers to pursue, in particular for materials that are going to help us live better.

**3** My favorite material is paint. In the last ten years the advances in technology in paint and surface materials have created many new design opportunities.

The way we cover surfaces is very important. How do we clean it? When does it have to be resurfaced? Unfortunately paint companies in the US look more at the mass market rather than at the individuals who are trying to do something to better the environments for which they are designing.

**4** Something very long-lasting and sustainable that requires hardly any maintenance. Because maintenance is our real problem.

## JOSEPH LEMBO

**1** I have been truly inspired by new materials I have seen while traveling in Japan, especially those used by fashion designers. Sometimes I am perplexed and cannot even begin to determine what the material or combination of materials I am looking at and/or touching may be. I am always impressed with the woven fabrics from Italy – such colors and unusual mixes of natural and synthetic fibers. I am always drawn to glass aggregate, paper and rocks on a beach.

**2** I am thrilled when a client is open to exploring innovative materials and is truly looking for something new. I still believe there is potential and new work to be found in packing, hospital, theatrical and obscure international materials that can

apply to commercial and residential use.

**3** I adore water, soaked cedar, and synthetically manipulated marble and granite chips and shavings.

**4** Transparency, strength, flexibility, lightness of weight.

## ARIK LEVY

**1** With materials I work like a scientist, I look for its genetic code to take apart and rebuild differently. It is subsets in our hands; even rigid materials conceptually become soft molecules to model and sculpt and identify.

**2** Innovation is a must, it is a priority, innovation is timeless, it goes across decades. With the introduction of new ways with materials, you obviously influence the production line, logistics and so on. What is important is the implementation of the idea. As much as new materials, innovation counts; to introduce it needs as much innovation.

**3** Intelligence – it is everywhere and makes lots of other ones. It has the most potential when it is used.

**4** My dream material is the one I dream of, and I wake up with it or to it and it will have imaginative properties.

## ROSS LOVEGROVE

**1** 'All my life I have enjoyed the reputation of being someone who disrupted prevailing ideas.' – Kepler

**2** It's only the future if it can't be made.

**3** More and more, my mind leans toward composites, the amalgam of fibers with resins. The polymerization of grasses and the textural beauty that results from such

collaboration will change our perception of rigid products from cars to aircraft to architecture. I am currently speculating on the potential of polystyrene, milled from form data and thin-skinned, as a way of provoking a new lightweight attitude to structure.

**4** A material that comes from the earth becomes something, then returns without trace. E.S.E. System: Earth, Sky, Earth. Proteins and Polysaccharides in Trinity.

## JAMES LUDWIG

**1** I think about materials in a much different way now that I'm a parent. What was cool before might now be seen as deadly. I read labels and want to know the contents. How long it will be around, and will it end up in breast milk or a class-action suit one day? I balance this with my fascination with the new.

**2** We approach materials in two ways: 1) force materials to address defined needs; 2) let needs be driven by material capabilities ... and on two levels: 1) material science – what's its impact on the user and the environment; 2) material sense – what qualities does it lend to the product and its environment?

**3** Wood – it's good dead or alive. It's unpredictable enough to keep it interesting.

Potential: alive – as a filter; dead – in high-tech ways with a low-tech material. It's a universally used open architecture.

**4** 'Skin': Self-healing; sensing; sealing, breathable, permeable; carries data; conducts electricity; controls temperature; turns colors based on environmental stimuli etc.

## GREG LYNN

**1** Materials are becoming more plastic with the use of manufacturing and prototyping that does not use tools or that uses inexpensive tools. A nose cone on an airplane can be pulled with incredible force over an epoxy tool that is shaped with a CNC machine: an incredibly expensive process that uses an incredibly inexpensive tool. The short runs and freedom of shape with materials like superformed titanium and aluminum make all kinds of new uses for materials as well as exotic shapes, textures and colors available for everything from products to architecture. The plasticity of materials also means that they no longer have as much propriety in terms of official uses. On the table, with Alessi, I am working with ceramics for flatware and titanium for a coffee service. The materials are used uniquely because the aerospace ceramic is as strong as metal but feels and tastes better for your mouth. I always start with the look and feel with materials and then try to design within their physical properties and within an economical limit to make things affordable and potentially popular.

**2** In architecture, it is more typical to start with a material and then begin the design from the material principles of construction and forming. This is the way we work, especially because of the model-building and prototyping equipment here in the office. The CNC router and the laser cutter are both robotic, as they run on code that we create on a computer. I tend to look for materials that can be formed and shaped with similar techniques that we use for studying construction using the 2-D laser cutter and the 2- and 3-D CNC mill. This has led me toward vacuum-formed plastics as an interior material, toward tooled fiber-reinforced resin and tooled MDF, and toward complex lattices of steel and wood components like space frames. I am a big fan of highly articulated and even decorative surfaces and structures, so I prefer to use a CNC machine to inscribe tooling lines on a surface instead of making it smooth. I like joints, edges, mating lines, multiple materials and components, so materials that are monolithic, like concrete, are not so interesting.

**3** Bent steel tubing at the scale of columns and beams. I am designing my own house right now, and it will be built out of radially bent steel tubes that twist like a pretzel to make the columns, beams and trusses all out of the same tube of steel. Powder-coated in a warm cosmetic color, I like to think of it as a resident of the Venice Beach custom-bike and -chopper culture at the scale of architecture. More and more, heavy metal is getting better and better. The mystique and heroism of Richard Serra are relaxing into a freedom and ease in rolling, extruding, bending, explosion-forming, pull-forming and superforming metals. I also am very excited about the merger of metals with ceramics that is going on in sintering. Once large-scale sintered components are available, architecture will have some great opportunities for new types of structural cladding. These are the materials that suggest an exoskeletal, lightweight architectural revolution in the spirit of Gaudí. Behrokh Khoshnevis is doing some provocative work on manufacturing houses using a technique he invented called contour crafting where concrete is CNC-trowelled in 3-D like a coiled pot and extruded to form buildings.

**4** 3-D woven textiles are a pet interest of mine as I have been visiting North Sails 3DL woven sail-makers in Nevada for years and have been trying to find a way to use a similar material in architecture. With Vitra we are working on 3-D woven upholstery that is very exciting. If we could weave large-scale 3-D textiles and stiffen them either with a matrix of plastic, as is done presently, or – even better – by coating and firing them as ceramics, I would be happy.

## SUSAN LYONS

**1** I wish I had paid more attention in chemistry class. Chemists are the chief architects of the materials we use, ingest and throw away every day. We need to recognize the enormous value of the materials that we are using and losing every day.

**2** Scientists, designers, manufacturers and business people have the collective power to transform how materials are deployed and re-deployed in our culture. I think true innovation will come when people from all of those disciplines are sitting around the same table.

**3** My favorite material is one that uses nature as a model – that happily and creatively resides in either a technical or a biological cycle as described by Michael Braungart and Bill McDonough in their book *Cradle to Cradle*.

**4** I dream of a material that is as smart, beautiful and nutritious as the leaf of a tree.

## CARL MAGNUSSON

**1** Materials are the latent museum that ideas must pass through to be realized, and pass they have throughout time; existing materials are to be revered and reinterpreted while new ones are welcomed into our vocabulary of form-speech.

**2** We use quite traditional materials such as steel, wood, aluminium, plastic etc. New materials really have to prove themselves. Like new ideas we are enchanted with new materials and want to understand their properties and potential uses that do not supplant existing materials. But we rise to new opportunities.

**3** Despite their age, aluminum alloys are continuously evolving into new potential usages. They can play traditional roles, as in the sand-casting of an Art Deco vacuum cleaner, and can be machined to be part of a motor or as a structural element on a racing car. Aluminum is traditional and modern at the same time. It recycles.

**4** Composite of wood and carbon fiber. Elastic metal for casting and extruding.

## FERN MALLIS

**1** Now that we have truly become a computer high-tech world ... much is done in a space we can't touch or feel. We are missing that 'touchy-feely' thing that makes us feel good.

Nothing compares to the way cashmere or certain silks or 800- or 1,000-thread cotton sheets on your bed make you feel. Also touching a highly polished steel wall, or a cool marble counter ... materials

bring back memories and make us think of new things to do with them ...

**2** In our world of fashion and fashion-show production, we are always looking for new materials for tents and tent liners, fabrics to cover walls with, waterproof and flameproof, and new materials for draping. For the clothing, it is about what can be washed, how garments can work and keep creases, or not, can absorb body heat, and keep one cool or warm all in the same day. The technology makes packing and traveling easier, and clothing more functional, comfortable, durable and either very affordable or very expensive ... In our world, everything starts with the textiles.

**3** Cashmere and paper. One makes me feel good, warm, sexy, touchable and expensive, and the other informs me and tells me what is going on in the world, and communicates to me and for me.

**4** One that can be woven into a garment and when you wear it you instantly lose ten pounds, and another that is more static, and one can just rub it and the genie pops out to grant me my wishes ... which are not for publication.

## EZIO MANZINI

**1** I spent the first part of my life as a researcher and designer trying to understand the new materials. I spent the second part of it trying to reduce their use. The turning point for me has been a better understanding of environmental issues and, in particular, of thinking in terms of sustainability.

Sustainability asks of us several radical changes. One of them is to dramatically reduce the ecological footprint of our way

of living and to dematerialize the systems of products and services on which they are based. In other words, we have to consume less material per unit of services that these systems ultimately deliver. And we have to do that in a radical way: to be sustainable, we people and communities in industrialized societies have to consume 90 per cent fewer environmental resources. In order for that goal to be achieved, we need new ways of living, new forms of organization and, as for what concerns us here, new families of materials.

**2** Considering the whole system of production and consumption, a dematerialization of the order of 90 per cent asks, of course, for less material. But it demands materials that be highly performing. The more we want to reduce materials, the more we have to pay attention to them and to their properties.

From the perspective of a sustainable society, we have to think of new-new materials: new materials with abilities that are different from the ones that characterized the old 'new materials' (i.e. the ones that predate sustainability consciousness).

**3** The 'grey material', i.e. our brain: the intelligence that is needed to learn how to live better consuming less and regenerating our living environments.

**4** I imagine two dream materials.

The eternal-material that lasts forever and that can continuously be recycled, giving form to different shapes and functions.

The instant-material that lasts just the time needed to deliver a service and, afterwards, disappears in some ecological cycle.

It must be observed that the biggest

problem for both, the one that makes it difficult for these dream materials to become real materials, is not a technological one. In fact, there are materials that, potentially, are very near to the dream materials but that, in practice, when integrated in their operative systems, do not work in this ecologically positive way.

The real problem is that potentially eternal and potentially instant materials ask for a high degree of 'care': somebody has to take care of them in every phase of their life, from when they become products to when these products are transported, used and, finally, dismissed. But, unfortunately, in the present day, care is the resource that appears to be the most scarce.

## PETER MARINO

**1** As well as the diversity of projects undertaken by the office, we are known for our application of materials in new ways. A residential client may love a certain furniture piece or local building material which may be the key to the entire project. Often, commercial fashion companies have proprietary colors and materials that we may study and reinterpret at an architectural scale. This approach is quite unique to this office and has been developed over many years of study and research into materials and manufacturing processes around the world.

**2** Materials are the building blocks of architecture and design; innovation in materials is innovation in architecture and design. Consequently our office accesses new materials and processes via the extensive network of manufacturers, artists and craftspeople who either bring new materials to us or respond to our specific requests to develop materials for particular project needs. Our latest building project, for example, the ten-story headquarters for Chanel Asia in Tokyo, is wrapped in a unique, programmable woven-glass, metal and light façade entirely custom-developed in response to one of the unique brand materials, tweed.

**3** Man-made materials have to be the answer as they are the future. The possibilities offered by plastics, composites and compound materials are endless in construction technology. Developments in glass fiber, carbon fiber and fiber-optic materials are providing entirely new building blocks to develop structures and finishes entirely impossible twenty-five years ago.

**4** Applied light; poured, painted, sprayed ... Architecture is inherently linked to space and light; just think about the possibilities of light available to be applied wherever and whenever required without any physical constraint of electricity.

## STEFANO MARZANO

**1** Materials are always a source of inspiration. At the same time, inspiration becomes a source for innovation in materials. We live in a physical world; anything non-physical has to go through some process to come into being such that we can interact with it. Material is a clear interface between us and whatever the world was, is and will be. And with each new exposure, you may encounter a material in a different form, which is inspiring.

**2** The creative process encounters challenges that force us to look for materials that stretch capabilities. In a way we close the loop – inspiration causes innovation as we look for materials that address new needs.

**3** I cannot disconnect material from application. I don't have a fascination for a specific material without a context. But I can be inspired by any material, by the historical connotations, by the way it sparks my imagination, its expressive elements, functionality and rationality. I love many materials. I love the materials that I can understand from the surface to the inside. These are the materials with which the code of communication and interaction with me is of a higher maturity level.

I am also intrigued by materials that I discover but don't understand: perhaps I can comprehend the surface but not the inside. I have been able to chop wood and break stone. But how do you come to know the inside of a composite? It is harder for those materials that are more distant from nature to communicate. They are a challenge because I have to work hard to understand them.

**4** I'm interested in the capabilities inherent in a material. Can we have a material that can have in itself all the possible properties known so far about materials? These properties can be brought into effect by a specific process. This material would be nothing by itself but would have in its DNA all properties. The material could be educated, could learn what it needs to, through the course of treatment.

**INGO MAURER**

❶ Most materials and things can be inspiring in one way or the other. It could be lips, tits, penne all'arrabbiata or the smell of 'Vol de Nuit' by Guerlain.
❷ Form follows the material's character. If you try to rape it, it will show. Understanding and developing a relation with the material is the first step to do.
❸ Paper, paper, paper! It creates soft, eye-pleasing light – which is my material to work with. I love ephemeral qualities, and I also love the great materials which are yet to be invented or discovered. My dream material would be 'snipping with the fingers – and there is light'. I shall work on this in my next life.

## ALBERTO MEDA

❶ The design process is not linear, it is not possible to plan it, it's rather a complex activity similar to a game's strategy, but the strange thing is that it's a game where rules are continuously changing, and for that very reason it is so fascinating and mysterious.

The designer collects his ideas from various sources, each in his own world of reference where he looks for creative suggestions.

I am personally interested in the world of technology and new materials because it seems to me to be the contemporary expression of the imaginative capability of man, of his ingeniousness fed by his scientific knowledge. Technology, in its hard and soft aspects, is, in my opinion, a tank of creative suggestions, because if we look at it with an interpretive glance, beyond its strictly technical performance, sometimes we get simple ideas which can

activate complex mental processes.
❷ Technology must be tamed in order to realize things that have the simplest possible relation with man, refusing the concept of technology-driven industrial goods without regard for human needs and with no communicative rationality.

Technology is not an end in itself but a means to produce simple things capable of enhancing expressively the space around them. It contains a paradox: the more technology is complex, the better it can produce objects with a simple, unitary, 'almost organic' image.
❸ I try to reduce the number of components through the integration of functions, because this matches the organic idea of having the same piece carry out several different functions. I think of new materials as those whose performances contribute to a renewal of the species of objects.
❹ A material able to change its properties and structure, depending on stress variations and external conditions.

## RICHARD MEIER

❶ One's always looking at materials in terms of what they can do and how they relate to ideas about architecture. I am always interested in finding new materials, or existing materials that I can use in a different way.
❷ In our work it is not so much materials themselves that we look at for innovation but rather how they are used. We take aluminum, a common material, and think about it as a panelized system: we are always figuring out how to do it better than we have done it in the past.
❸ My favorite material is paint. I see the

most potential in materials that are economical and durable.
❹ My dream material would be seamless, able to be used on any kind of surface, suitable for interior or exterior use, malleable, inexpensive and white.

## PETER H. MEYER

❶ Material, mind and perception are locked together in a unique, singular relationship: can any one of these exist without the other two? Of course one can imagine materials existing in a cosmos without living creatures, without mind and perception. But perceiving the different qualities of the different materials, utilizing them by the power of human creativity for the purposes they fit best, transforming them deliberately, this is one foundation of human civilization – a never-ending process of innovation in the use of materials.
❷ 'Concentration on the essential' is the essence of our strategy; we are focused on innovative products, people and communication, stripping away all the meaningless glamour of the conventional. If there wouldn't be a steady flow of new ideas, of striving for better solutions for the future, my company would simply not exist.
❸ I admire the sheer endless potential of polymers. They allow so many applications, approach us in so many extremely different ways, it is incredible. If archaeologists of the present talk about the Iron Age or Bronze Age, archaeologists of the future may name our era the Polymer Age – for me without having any idea what may follow after this.

But when I really look deep into myself, beyond the fascination of the ever-changing face of polymers, natural wood is the closest to my heart, has the strongest emotional impact on me. Working with the furniture business for more than thirty years obviously leaves its traces. From my very personal point of view, no material can ever compete with the beauty and charms of a carefully designed and skilfully crafted piece of furniture made of selected, genuine woods.

4. There can never be a single 'dream material' for me. Materials have to fit to their purposes and the needs of their individual applications. The incredibly wide range of polymers is beyond all imagination. It really could become a dream the more this diversity reaches to a synthesis with sustainability and the ecological use of the resources of our world.

## ISSEY MIYAKE

1. It is the relationship between diverse materials and the human body that interests me.
2. Our current project, A-POC, is an innovative method by which a piece of thread goes into one end of a machine, and, in a single process, fabric, texture and a garment are created.
3. The greatest potential lies in making things that will serve people's needs in the future.
4. Nature.

## KARDASH ONNIG

1. I'm interested in how materials can have a spiritual quality through their connection to us. We use them in a certain place, in a certain way. An animal becomes leather; there's a kind of sacred deal you made with the material while it was still alive that you honor through using each different part of the former being. There are qualities of ritual, transformation, attachment and evocation in the process.
2. For me the innovation is in the transformation from one state to another. I'm interested in a single material, like leather, from which I can make shoes, toys, drums, lots of products out of it that serve vastly different purposes. We can expand the dimensions of the materials in terms of what they can do.
3. Stone, the hardest thing to carve. I find it difficult within my own work to make it spiritual, to bring emotional qualities to it, because it's hard to communicate with. It's uncontrollable. It's ten million years old! It's not easy to tame. But the acoustic properties of stone are like a musical instrument. It's the one way that stone is soft.
4. Crystal that you can carve. Every material has its vibration that you feel when you carve. It's something you can feel. What would the spiritual relationship be between you and the crystal? I'm looking for the ultimate mystical experience of the craftsman working with a material.

Also, what would it mean for the technology to change so that materials can be personalized? Otherwise what else do we have? The material should make a connection to the individual human spirit.

## GAETANO PESCE

1. Materials inspire me 40 per cent.
2. Not to repeat and to be innovative.
3. The most potential are the materials of our time.
4. Gas with mechanical capacity.

## ANDREÉ PUTMAN

1. My range of interest goes from the poorest American Olean to the most magnificent gold-plated mosaics – in homeopathic quantities. I like to marry them.
2. After engaging the rich to the poor, I like to play with finishes like a matte pale stone with satin, or rough slate with shiny wood. I am also a specialist in marrying the old and the young, like lava and Corian®.
3. There are two: enameled lava, which I use in very odd places and unexpected colors, and a magnificent fabric I designed for Maharam: fake horsehair, which is both technological and reassuring, comfortable and practical. If you upholster the walls of a room with it, you'll feel like inside a bronze cube, but we will also make sneakers with it! The possibilities are unlimited.
4. It would be a material which cannot get dirty. If not, one could wash it with a hose. There are many times when I wish I would have a hydrant in my home. I would hook up a hose and flush down all the dust.

## KARIM RASHID

**1** I see materials as inspiring new human behaviors, insuring the development or use of new materials. I am a great believer in the casual engineering of our world. The new relaxed world and casual conditions of lifestyle are not just relegated to form, shape and behavior but also refer to our new material landscape where tactile surfaces give us comfortable, engaging physical experiences. New polymers such as synthetic rubbers, santoprenes, evoprenes, polyolefins, silicones and rubberized coatings all contribute to this new softness. Multiple materials on a singular object by new technologies such as dual-durometer molding, co- and triple injection enrich the interface with our bodies. We have heightened our experiences via touch. Materials can now flex, change, morph, shift color, cool and heat etc. due to the 'smart'-material movement. From toothbrushes to color-changing hot wheels, these new, stimulating phenomena all play a role in our new relaxed environments.

This Casualization of Shape, form, material and behavior is definitely a movement – and a movement that is American. America is, to use a Victor Papanek term, casually engineering the rest of the world. Yes, America has created a kind of digital language of blobifying our world, physically and materially. This is not the first time America has literally shaped the world. The blobject movement is here thanks to a handful of designers such as myself who have taken a great interest in organic form and the technology that is allowing us to morph, undulate, twist, torque, blend and metaball our concepts. Also, new production processes and methods are contributing to this New World language.

**2** 'Smart' materials. Another sort of precision and convenience is offered by smart materials like polymer resins. While we expect material to change due to weathering and age, smart materials anticipate variance – they can mutate into a desired state under certain conditions. For example, thermoplastic mouthpieces for athletes undergo molecular change when heated. To achieve a perfect fit, you submerge the device in boiling water and then hold it in your mouth for a few minutes while it conforms to the exact contours of your teeth and gums as it cools. Another industrial material, nitinol wire, an alloy of nickel and titanium, has shape-memory. No matter how bent out of shape it becomes, nitinol will assume its original form when exposed to elevated temperatures. So far, its applications include a showerhead valve that automatically cuts off when the water gets too hot and wire-frame eye-glasses that when bent can be easily restored to their original form by immersion in hot water.

**3** I like and work with almost all materials you can think of. I rarely work with wood, but I recently completed a restaurant for Morimoto the Iron Chef in Philadelphia that is almost all bamboo and plaster (all natural materials). My favorite materials are ones that can be almost immaterial in appearance – glass or plastics. Plastics provide the most varied performances, finishes and diverse properties. They inspire me to create fluid, soft, organic, almost human form.

**4** It is anticipated that nanotechnology will have a momentous impact on variance. By arranging a material molecule by molecule, atom by atom, the atom can become a modular building block. Developed through micro-machines, using gears, levers and switches of only two to eight atoms, such designs will be able to be self-contained, self-propelled, programmable, intelligent and regenerative. Reassembling materials will be triggered by certain amounts of electrical current, light, heat and pressure. We may well find ourselves one day in the world of object-morphing in response to just a touch or a small electrical charge.

## IVY ROSS

**1** I'm in love with texture, color and form. You become a co-creator with the material; what inspires me is that you take the material and collaborate with it to bring it into a new form. It gives you a springboard to be a truly creative person. The material is incomplete without you playing an active role with it.

**2** Reading about new materials gives my brain a jumpstart. It gets my mind fantasizing and applying that fantasy either to that particular material or to a different one. I like to look at the way something is constructed and apply it to something completely different. Materials have so much information to give you. You can use them literally, or as information that leads to fantasy or inspiration. There's an act of connection and translation. They act as a catalyst.

**3** I love a recycled solid-surface material called Alkemi®, which uses a salvaged aluminum waste, fillers and pigments in a monomer base. I love recycling. Material creation is in part like alchemy, it's about creating something from nothing. I've been particularly fascinated by materials that are made of recycled materials.

Another favorite is Krystalweave, which is made of clear extruded-polymer yarn. A woven cloth, it gives the feeling of light and translucency but has enough weight to create a spatial division.

There's a theme here: I'm intrigued by things that were designed for one kind of fabrication and are used for something else.

**4** To be able to be in conversation with me and my thoughts. It could change and adapt and react to my creative thoughts.

## LEE SKOLNICK

**1** Materials and their evocative qualities are of particular interest to our firm. We are concerned with telling stories that are expressed through architecture and design. Materials become a language for storytelling, a way for us to express our ideas. We are drawn to a given material's ability to evoke associations, as well as its individual qualities (color, texture, reflectivity).

**2** Even the nature of the source of a material and the process it went through during manufacture become part of this expressive quality that we attempt to mine and use in our designs.

**3** Without having a specific favorite material, we are concerned with the environmental qualities of materials, in terms of both their manufacture and how they react in a given space. We are always looking for materials that we can be

satisfied are environmentally friendly throughout their entire life cycle.

**4** The richness and the experience that you can create with materials is in their juxtaposition. I am drawn to the sensuous qualities of naturals and synthetics. At the same time we have a particular interest in materials that can carry some sort of information. Not just print or visual information but something transformative. For example, lenticular wall coverings allow us to create an environment that changes as you move through it. Materials which have that potential are very attractive to me.

## ALISON SPEAR

**1** Materials always are an inspiration in my projects. I like glamour, I think that materials are like fine clothing that dresses and enhances the structure beneath. Sometimes it is just the material that drives an idea or concept. In architecture, design and fashion, materials are the tactile and visual elements that make a project 'sing'.

**2** I like to use materials in an unexpected way ... one of my favorite challenges is to find a material that is unintended for its original use and to use it in a fun and innovative way – i.e. rubber gaskets could be finials ... acoustic panels as wall coverings ... decorative meshes as wall partitions ... glass used structurally ... I also like to find inexpensive things and use them in an expensive way ... I see beauty in industrial and 'found' objects.

**3** My favorite materials are the ones that I am working with at the moment ... all of a sudden golds and bronzes look rich and beautiful to me ... sometimes it is a color

that gives me a buzz, so I look for everything and anything in that color, rubber, leather, cork, rope, metals, fabrics, glass ... The other day I was wandering through Home Depot and a local hardware store to give me ideas for materials to help inspire a new design project – in this case an upscale hot-dog stand on Miami Beach! By the time I was finished, I had a bagful of rat-traps, washboards, construction lights, masonite, red rubber-tipped clips, carabiner parts, rope and a set of dominoes in a wood box!

**4** My personal dream materials are ones that are irresistible to touch and see: a velvety stone, thick 3-ply cashmere, alpaca, fine leather, smooth silk, waxed wood. Things that you can't take your hands or eyes off! I like to touch things – materials should be inviting, intriguing and romantic.

## PHILIPPE STARCK

**1** There is, then, in front of us, the same subtlety of meanings, the same multiplicity, the same indefinite number of meanings that a collection of letters and musical notes has. It is fundamental not to forget that all these are here to give only sense, and that only the meaning of the story is interesting, and only if the story is interesting for the people who will read it, only in this case does the story deserve to exist, and as a result the object is worthy of existing, and then the material is worthy of existing. What I mean by that is that the material does not matter, its color does not matter, except what it is going to bring to the person who is going to live with it; because the final goal is the human being, it is us. There is no material that is worthy of existing if it does not have any benefit, whatever it will be, for human nature. As a result, my answer would be that materials, whatever they are, colors, no matter which one, I do not care at all about them, as I am only interested in the final result. Materials are vehicles, a way to talk about certain things – and those certain things end up being relations with tenderness, love, dream, vision, mutation, humor and poetry.

**2** Innovation is extremely important in my work, because as I just explained before, each new material is a new word, an additional word, which allows you to bring life to a new concept. So each new material, each new innovative material, is a possibility to tell a new story or to tell more precisely the same story, or to make more accessible a story which was before told only to certain people because of the uniqueness or cost of the end product. So, of course, innovation is fundamental; it is the richness of the meaning, the richness for the human who is going to live with it.

**3** My favorite materials have always been synthetics for various reasons. First, synthetics are structurally admirable because they are the fruit of the human intelligence. They were invented by human nature. I do not believe in God at all, but to summarize my thought, I am not interested in using materials that are coming from nature since they are supposed to have been created by God. Of course, this explanation sounds a little bit extreme, but it is to emphasize even more how fond I am of human creation, a fanatic of human intelligence and of all its byproducts. Synthetics, coming from human intelligence, are nowadays

superior to natural materials. Nowadays the extraordinary capabilities of some synthetic materials go beyond the capabilities of some natural materials. So we went beyond the status of natural, we went beyond the status of God.

Outside of these neo-philosophical reasons, synthetics are the only ones that adapt to our mutations and the necessity of mutation that our societies require. Only plastics can be the permanent accomplice of our mutation. Thirdly, only synthetic materials allow us to sow the quality of a product while reducing its price and to multiply it to infinity in order to create a higher-quality product for a larger number of people. Synthetics are necessary to democratic design, in other words to the democratization of good ideas – in other words, the only thing I am interested in.

**4** I do not really have a lot of imagination. It is extremely hard for me to dream of a new material for the good reason that what I believe is already in preparation seems to me already exceptionally interesting and surprising. When you are already able to see materials that are self-healing, it is terrific. When we are able to witness the first steps of programmable materials, which will be able to take any kind of form or function depending on the order we give them, we are in a quasi-absolute dream.

To finish, my dream material is an immaterial one, because I only believe in the future of dematerialization by definition. Our mutating civilization is going essentially toward dematerialization because it is an instinctive obligation to know that life was created four billion years ago and will disappear with the solar system implosion four billion years

from now. We are now in the middle and have four billion years to become – ourselves as well as everything surrounding us – totally immaterial, as we will have then to move away. So, nowadays, the most interesting materials are those that are 'dematerializable'.

## PETER TESTA

**1** New matter is creative. Materials are not simply passive objects to be manipulated by active subjects. Just like nature itself, materials are hybrids of biological, chemical and human processes that are increasingly open to design.

**2** To creatively engage the virtualities of materials, our office operates at the shifting nexus of architecture, advanced engineering, material science, artificial intelligence, manufacturing and process engineering. Through new materials we imagine emergent possibilities for form and space as they give durable shape to schemes, forms and relations that are not already present in another form in other materials.

**3** What interests me is discovering the propensity of materials to invent structures and forms. I see great potential in fibers, foils, films and foams. For example, our firm is currently developing Carbon Tower, a high-rise made entirely of woven and nonwoven composite materials and thin films. But we also work with many other materials such as pulp and waste-paper products.

**4** My dream material is MoSS, an active material system we have developed to evolve multi-cellular structures at all scales. A vital property of MoSS is that it can initiate emergent, self-regulatory patterns wherein structural morphology and material properties search for each other and co-evolve within a dynamic shaping environment.

## TUCKER VIEMEISTER

**1** Objects Matter. That makes designers very important. As Adrian Forty wrote in *Objects of Desire*, 'Designers have the capacity to cast myths into an enduring, solid and tangible form, so that they seem to be reality itself.' That's because reality is made of real things, everything else is only figments of our imagination.

I design real things – that means they have to be made out of some kind of materials. Life may be 'like a play', as Shakespeare said, but we really need props in order to make it meaningful.

As an industrial designer, I usually skip the step of selecting a material before I design something – because either the client already has a machine that uses a certain material or because I assume that I can find a suitable material with just the right properties somewhere – and of course if I can't find the perfect material, then that constraint will inform the evolution of the design – and make it better. (See, using this strategy, I always win!)

All that said – I love the stuff things are made out of, and I love to see it – not disguise it. I love to see the shine or the grain. I love to see the materials age, scratch and patinate over time.

**2** Before now, technical innovation drove the design process. When stones were invented, cavemen could make stone tools and weapons. When computers were invented, we could do all kinds of new things, like make animated banner ads etc. Now technology has caught up with our imaginations – scientists can make almost anything designers can dream of. This makes our new age totally different than anything before.

It puts a lot of responsibility on designers, manufacturers and businesses, who in the past were basically the same as cavemen, just trying to survive in the battle against nature. Now designers need to lead. Instead of just translating technology into useful things and feeding the business machine, instead of figuring out what customers WANT, we must decide what we NEED.

**3** Gold. It's shiny. You can hammer it or cast it. You can wear it. It's valuable by itself – you can make money out of it or a bowl. It's beautiful. It's soft, warm and golden. I've had gold fever for a long time.

**4** I'd love to have a magic material that I could work in my hands like clay – but that I could adjust the properties of like a Photoshop file. First it would be soft so I could push it around, then hard so I could sand it, then finally permanently rubbery! By clicking a box or moving a slider – I could make it a liquid or clear or hard or soft or metal or rubber. Then, of course, it would be easy to email it or 'print' duplicates with little cost and 'melt' it down to recycle into the next thing, so it would have little adverse environmental impact – but huge cultural impact! It would allow everyone to be a designer. We would all be rich and happy!

## MASSIMO VIGNELLI

**1** In our design process, materials play a dominant role, therefore the necessity of sifting the unlimited offering into a limited range, to form a vocabulary that can better express our intent. This is a process that resists easy temptations in favor of long-lasting choices. The right material is the right choice.
**2** Innovative materials and processes go together and are used with appropriateness to the project at hand, to better express the nature of the material for that particular application.
**3** The one that is the most appropriate for the project at hand. However, I like to have a range of materials that are apt to better express our language.
**4** I have no dream material. Each project has its own requirements, and one should listen to them and creatively transform them according to one's own language and sensibility.

## KEVIN WALZ

**1** Almost always materials are at the forefront of my thinking. I'm always looking at a material and thinking what is the potential inherent within the material and wondering what properties it has that haven't already been exploited. I'm asking myself what I can do with something new. For example, with a combination of carbon fiber and wood as a laminate I developed a collection of furniture, including a dinner chair that weighs a pound. It's like plywood for the next century; it consumes less wood but has the same richness, and can be used more like steel in that one can thermoform it. It's rigid and resilient.

**2** I think the idea of innovation has been confused with flashy forms and humor, particularly in the last few years. There can be an idea of making a visual joke or creating a visual pun with imagery which really isn't about innovation. Innovation comes from the harnessing of a process or a material. Sometimes I'm attracted to a material that is essentially natural and that we can manipulate or do something to because of new technology. Sometimes I can't find what I need, and I have to develop a new process or borrow from another industry. It has to have a meaning for me.

**3** I don't have a favorite. Some days I love cork best: it's an isolator of sound and motion, it's anti-bacterial and resilient, it has perfect memory. It's an industrial material that has been forgotten. It's also naturally ergonomic.

Where I would love to see major change is in housing, because there is so much wasted material.

**4** I've been very intrigued with the idea of thinness, so the laminates I've been working on excite me. I like the idea of carbon fiber because it's really a fabric which obtains a strength and a structure when it's resined. One of the ways I always measure a material is to ask how it is going to age. Naturals tend to age well. There are new materals – resins, for example – that age beautifully too. I hate gypsum board and plastic.

**VICENTE WOLF**

**1** My wish is that materials look as delicate as rose petals and are as tough as iron, smooth as silk but able to withstand long wear. Fabrics that are able to alter color depending on light, materials that are flexible to mold to unusual shapes, weaves that if punctured can reseal themselves. Surfaces that can transition from clear to opaque and from liquid to solid.

**2** The role is giving me a broader pallet of elements to work with in creating functional, attractive and inviting environments.

**3** Concrete that is able to transform itself from industrial to high-end market by altering colors, texture and design for residential and commercial spaces.

**4** Material that never gets dirty, walls that never dent, floors that never scratch or need cleaning, paint that is able to alter its color without having to be repainted.

Addington, D., Michelle Daniel, and L. Schodek, *Smart Materials and Technologies in Architecture* (Burlington: Architectural Press [Elsevier], 2004)

Ashby, Michael, and Kara Johnson, *Materials and Design: The Art and Science of Material Selection in Product Design* (Oxford: Butterworth-Heinemann, 2002)

Askeland, Donald R., and Pradeep P. Phulé, *The Science and Engineering of Materials*, 4th edn (Salt Lake City: Thomson-Engineering, 2002)

Benyus, Janine M., *Biomimicry* (New York: Perennial, 1997)

Callister, William D., *Materials Science and Engineering: An Introduction*, 6th edn (New York: Wiley, 2002)

Gay, Daniel, Suong Van Hoa, and Stephen W. Tsai, *Composite Materials: Design and Applications*, 4th edn (Boca Raton: CRC Press, 2002)

Hongu, Tatsuya, and Glyn O. Philips, *New Fibers*, 2nd edn (Cambridge: Technomic Publishing Company, 1997)

Hull, D., and T. W. Clyne, *An Introduction to Composite Materials*, 2nd edn, Cambridge Solid State Science Series (Cambridge: Cambridge University Press, 1996)

Kadolph, Sara J., and Anna L. Langford, *Textiles*, 9th edn (Upper Saddle River: Prentice Hall, 2001)

Kalpakjian, Serope, and Steven R. Schmid, *Manufacturing Engineering and Technology*, 4th edn (Upper Saddle River: Prentice Hall, 2000)

Kingery, W. David, H. K. Bowen, and Donald R. Uhlmann, *Introduction to Ceramics*, 2nd edn (New York: Wiley-Interscience, 1976)

Lupton, Ellen, Jennifer Tobias, Alicia Imperiale, Grace Jeffers, and Randi Mates, *Skin: Surface, Substance, and Design* (New York: Princeton Architectural Press, 2002)

McDonough, William, and Michael Braungart, *Cradle to Cradle: Remaking the Way We Make Things* (New York: North Point Press, 2002)

Mori, Toshiko, *Immaterial/Ultramaterial: Architecture, Design, and Materials (Millennium Matters)* (New York: George Braziller, 2002)

O'Mahony, Marie, and Sarah E. Braddock, *Sportstech: Revolutionary Fabrics, Fashion, and Design* (London: Thames and Hudson, 2002)

Satas, Donatas, Arthur Tracton, and Arthur A. Tracton, *Coatings Technology Handbook*, 2nd edn (New York: Marcel Dekker, 2000)

Wessel, James K. (ed.), *The Handbook of Advanced Materials: Enabling New Designs* (New York: Wiley-Interscience, 2004)

Wigginton, Michael, and Jude Harris, *Intelligent Skins* (Burlington: Architectural Press, 2002)

## A

### ABACA
Also known as manilla hemp, abaca is the strongest known vegetable fiber and is obtained from the leaves of a tree in the banana family. The fibers are between 4 and 8 ft/1.2 and 2.4 m long, lightweight, soft, lustrous and nearly white. They do not swell or lose strength when wet. Denier ranges from 300 to 500.

### ACETAL
An engineering thermoplastic produced by the polymerization of purified formaldehyde ($CH_2O$). It has a high melting point, high strength, good frictional properties and resistance to fatigue. Industrial end-users are familiar with acetals in the form of gears, bearings, bushings, cams, housings, conveyors and any number of moving parts in appliances, business machines etc.

### ACRYLIC
Polymethylmethacrylate (PMMA) has excellent clarity and ultraviolet resistance, good abrasion resistance, hardness and stiffness, and low water absorption and smoke emission. It does, however, have poor solvent resistance and a low continuous-use temperature of approximately 120°F/50°C. Its uses include lenses, light covers, glazing (particularly in aircraft), light pipes, meter covers, bathroom fittings, outdoor signs, skylights, baths and toys.

### ACRYLO-NITRILE BUTADIENE STYRENE (ABS)
An engineering thermoplastic that is hard and tough, though it has poor solvent and fatigueresistance. It is typically used for computer cases and domestic appliances and, in a blended form with polycarbonate, for automotive interior surfaces.

### ACTUATOR
A device that performs a mechanical action in response to an input signal, which may be either electric or fluidic.

### AEROGEL
An extremely lightweight, highly porous foam with a high surface area and very low density that can be made of silica ($SIO_2$), alumina ($AL_2O_3$), carbon and other materials.

### AGAVE
An American tropical plant with basal rosettes of fibrous sword-shaped leaves.

### AGGLUTINATES
Particles held together in a solid mass.

### AGGREGATE
A granular material, such as sand, gravel, crushed stone and iron blast-furnace slag, used with a cementing medium to form a hydraulic-cement concrete or mortar.

### ALCRYN®
A thermoplastic elastomer manufactured by Dupont that is melt-processible and has the properties of a vulcanized rubber.

### AMORPHOUS
A material having no defined crystal structure. Also known as a glass-like or glassy structure.

### ANILINE-DYED (LEATHER)
A leather that has been dyed by immersion in a dye-bath and has not received any coating or pigment finish.

### ANNEALING
The slow heating or cooling of a metal or glass in order to alter its properties.

### ANODIZING
The formation on the surface of certain metals, particularly aluminum and magnesium, of hard, stable surface coatings (an oxide of the metal) that have good electrical-insulating properties and can absorb dyes and pigments, allowing finishes to be obtained in all colors including black. The coating is deposited on the metal via an electrochemical process in which the metal is the anode; hence it is called an anodic coating, and the deposition process is called anodizing.

### ARAMID
The generic name for a class of flame-retardant polyamide fibers. Some types are used for protective clothing, tire cord and bullet-resistant materials – e.g. Dupont's Kevlar® and Nomex®.

## B

### BAGASSE
The dry, fibrous residue remaining after the extraction of juice from the crushed stalks of sugar cane, used as a source of cellulose for some paper products.

### BALLISTIC MATERIAL
A material developed for protection against projectiles such as bullets and small missiles.

### BINDER
An organic or inorganic substance added to particles of a material to hold them together.

### BIOCOMPOSITE
A composite material made from the combination of two or more natural materials.

### BIODEGRADABLE
A material that can be decomposed by microorganisms in the soil, as well as other natural factors such as weather, plants and animals.

### BIOPOLYMER
A biopolymer is a polymer found in nature or polymerized from natural raw materials. Starch, proteins and peptides, and DNA and RNA, are all examples of biopolymers in which the monomer units are sugars, amino acids and nucleic acids, respectively.

### BOUCLÉ WEAVING
A process that creates a yarn with loops, producing a woven or knitted fabric with a rough appearance.

## C

**CALENDERING**
A continuous method for producing a plastic sheet with a smooth finish and a specified thickness and width by passing the raw materials through a series of pressure rollers.

**CAPROLACTUM**
Chemical used as a feedstock for the production of nylon 6,6.

**CARBON FIBER**
High tensile-strength fibers between 7 and 8 μ/70,000 and 80,000 Å in diameter, for reinforcing composites, produced by decomposing with heat (pyrolysis), in an inert environment, such organic fibers as rayon, polyacrylonitrile (PAN) or petroleum pitch. Carbon fibers contain between 93 and 95 per cent elemental carbon.

**CELLULOSE ESTER**
A family of polymers that includes cellulose acetate. They are derived from naturally occurring cellulose (from cotton or wood) and are widely used in the fabrication of membranes and as the base material for some types of photographic film.

**CHEMICAL-VAPOR DEPOSITION (CVD)**
A coating process in which the coating is formed as a result of a chemical reaction between gaseous reactants at elevated temperatures in the vicinity of the substrate. It is used to deposit thin, hard films for semiconductors or protective coatings.

**CO-EXTRUSION**
Several layers of material are combined using a die designed with multiple flow-distribution channels, so that the extruded layers come together after they have been evenly distributed initially, thereby creating unique product properties compared with monolayer materials.

**COLD ROLLING**
A process that involves rolling metal at a temperature below its softening point to create strain-hardening (work-hardening). Cold rolling changes the metal's mechanical properties and produces certain useful combinations of hardness, strength, stiffness, ductility and other characteristics known as tempers.

**COMPOSTABLE**
A product that can be placed into a composition of decaying biodegradable materials and eventually turn into a nutrient-rich material.

**COMPOSITE**
A material created from a fiber (or reinforcement) and an appropriate matrix material in order to maximize specific performance properties. The constituents retain their identities; they can be physically identified and exhibit an interface between one another.

**CONSTRAINED-LAYER DAMPING**
This is a 'sandwich' construction material consisting of a visco-elastic material bonded with a relatively stiff constraining layer. When the system flexes during vibrational distortion, sheer forces are created on the stiff constraining layer, and the vibrational energy is dissipated through sheer deformation (tensional forces).

**CROCK-RESISTANCE**
The ability of a printed surface to withstand repeated rubbing with another printed surface.

## D

**DENIER**
A unit for expressing the linear density or fineness of a filament, fiber or yarn. It is based on the weight in grams of 9,000 m – i.e. if 9,000 m of a fiber weigh 1 g, the fiber is 1 denier (1 den). Sheer hosiery yarn is 15 to 20 denier; the lower the denier, the finer the yarn.

**DEVITRIFICATION**
The process by which a glass (noncrystalline or vitreous solid) is transformed into a crystalline solid. Devitrification may also occur on the surface as a result of the unstable composition of the glass or of unsuccessful annealing or accidental firing.

**DICHROIC**
Glass or plastic that has been coated with a thin layer of metallic oxide. Dichroic coatings transmit certain wavelengths of light while reflecting others, creating an interference effect similar to iridescence.

**DIELECTRIC STRENGTH**
An index of the ability of an insulating material to reduce the transmission of an electrostatic force between two charged bodies. The lower the value, the greater the reduction. A good dielectric is a good insulator, but the reverse is not necessarily true. Dielectric-constant values decrease with increasing temperature. Some typical dielectric constants at room temperature are: air 1; polyethylene 2.3; nylon 3.0 to 3.5; diamond 5.7; glass 7.0 to 7.6; water 78.54.

**DIMENSIONAL STRENGTH**
A measure of the change in dimensions of an object as the result of temperature, humidity, chemical treatment, aging or stress. It is usually expressed as a change in the units of the dimension per unit.

**DYE-SUBLIMATION**
A printing process in which ink is applied by the direct change of a solid to a vapor without becoming a liquid.

**E**

**E-GLASS**
A glass fiber (of calcium aluminoborosilicate composition) with high electrical resistivity that is used as a reinforcement in plastics and for electrical laminates. Also called electric glass.

**ELASTOMER**
An elastic, rubber-like material that recovers approximately its original shape and size after being stretched. Elastomers include vulcanized natural rubber as well as synthetic rubbers – e.g. thermosets such as neoprene (polychloroprene) and thermoplastic polyolefin (TPO) rubbers.

**ELECTROCHEMICAL DEPOSITION**
The chemical deposition of a thin layer of metal onto a substrate by means of an electrical current.

**ELECTROMAGNETIC INTERFERENCE (EMI)**
Undesirable interference, particularly in electronic devices and equipment, from radiation from sources such as other electronic devices, electric motors and other sources of electromagnetic energy. Most interference problems are in the radio frequency range (RF) range from 10 kilohertz (KHz) to 100 gigahertz (GHz).

**ELECTROSTATIC DISCHARGE (ESD)**
Static charge which builds up usually as a consequence of rubbing effects; it can generate voltages as high as 30 KV. When it discharges, ESD can destroy electronic components.

**EPOXY RESIN**
This thermosetting polymer resin characterized by epoxide molecule groups is used chiefly in strong adhesives, coatings and laminates. Epoxy resins have excellent mechanical properties and good dimensional stability.

**ESTER**
An organic compound that corresponds in structure to an inorganic salt. Esters are considered as being derived from organic acids. For example, acetates are esters of acetic acid.

**ETHYLENE-TETRAFLUOROETHYLENE (ETFE)**
Also known as Teflon® (the brand name for Dupont's ETFE resin), this thermoplastic member of the fluoropolymer family is noted for exceptional chemical resistance, toughness and abrasion resistance.

**ETHYLENE VINYL ACETATE (EVA)**
A thermoplastic resin typically used as stretch film for shrink-wrapping and as shoe soles (often as a foam), disposable medical equipment, flexible toys, tubing and wire coating. EVA's are also used in many hot-melt manufacturing processes such as packaging, bookbinding or label-sticking.

**F**

**FEEDSTOCK**
Raw material supplied to a machine or processing plant from which other products can be made.

**FELT**
A nonwoven, fibrous textile composed of fibers interlocked by mechanical or chemical action, moisture or heat.

**FILAMENT**
A continuous fiber for use as a reinforcement, usually made by extrusion from a spinneret. Filaments are typically extremely long with a very small diameter, generally less than 25 μ/1 mil.

**FLAX**
The fiber of the flax plant that is made into thread and woven into linen fabric.

**FLUOROCARBON**
An organic compound containing fluorine directly bonded to carbon. Fluorocarbons are the basis for the production of fluoropolymer resins such as Teflon® and exhibit good resistance to corrosive environments, as well as stability at elevated temperatures.

**FLOAT GLASS**
Glass formed by a process of floating the material on a bed of molten metal. Without polishing and grinding it produces a high optical-quality glass with parallel surfaces.

**FLY ASH**
The very fine ash produced by the combustion of coal or coke, a mixture of aluminum oxide, silicon dioxide, unburned carbon and various other metallic oxides. When recovered, it can be used as a filler for cement or plastics and also as a constituent of some commercial products.

**FORMALDEHYDE**
A gaseous chemical used in the manufacture of melamine, polyacetal and phenolic resins; for durable-press treatment of textiles; and to make foam insulation, particleboard and plywood. It is a carcinogen that is toxic by inhalation and a strong irritant.

**FRP**
Fiber-reinforced plastic. Also known as GFRP (glass fiber-reinforced plastic).

**G**

**GEL**
A colloid in which the dispersed phase (particles between 1 nm and 1 μ) has combined with the continuous-liquid phase to form a viscous jelly-like product.

**GEOPOLYMER**
A material formed by polymerization that exhibits the most ideal properties of rock-forming elements – i.e. hardness, chemical stability and longevity equal to natural geological longevity. Examples include alumino-silicate compounds.

**GEO-TEXTILE**
A synthetic fabric used to retain soil while allowing water to pass through both soil and fabric into a drainage-collection system.

**GLASS-CERAMIC**
A devitrified or crystallized form of glass with properties that can be varied over a wide range.

**GYPSUM**
A colorless mineral (calcium sulfate with attached water molecules) used to make plaster of Paris and various plaster products.

**HEMP**
A coarse, durable bast fiber obtained from the inner bark of the hemp plant. Used primarily in twines, cordages and apparel.

**HIGH-PRESSURE LAMINATE (HPL)**
Materials manufactured from layers of fibrous material (e.g. paper) are impregnated with thermosetting resins and bonded by heat at a pressure not lower than 7 Mpa; one or both external surfaces are colored or decoratively patterned.

**HOMOPOLYMER**
A polymer resulting from polymerization of only one kind of monomer.

**HOT-DIPPING PROCESS**
The coating of a surface by dipping or immersion into a series of cleaning and pre-treatment chemicals prior to immersion in a molten liquid used as a coating. A typical example would be the galvanizing of steel with zinc.

**HYDRAULIC CEMENT**
A mixture of finely ground lime, silica and alumina that sets to a hard cement when mixed with water.

**HYDROPHILIC**
Having an affinity for water, thus readily absorbs or dissolves in water.

**HYDROPHOBIC**
Repelling water, thus does not combine with or dissolve in water.

**HYDROXYAPATITE**
This ceramic, the chief structural component of bone, is composed primarily of calcium phosphate crystals.

**HYPALON®**
The brand name for a chlorosulfonated polyethylene manufactured by Dupont Dow.

**HYTREL®**
The brand name for a range of polyester elastomer engineering polymers manufactured by Dupont.

**I**

**INCONEL®**
The trade name for a series of corrosion-resistant alloys of nickel chromium and iron manufactured by Inco.

**INJECTION-MOLDING (CERAMICS AND METALS)**
Parts are formed by heating a mixture of ceramic or metal powder plus a binder, then forcing the softened material into a closed die, processing the formed part to remove the binder and densifying it by means of high-temperature sintering.

**INJECTION-MOLDING (THERMOPLASTICS)**
The melted plastic is injected into a mold cavity, where it cools and takes the cavity's shape. Details such as screw threads and ribs can be integrated, thus allowing the molding operation to produce a finished part in one step.

**INJECTION-MOLDING (THERMOSETS)**
The thermoset material is first heated until it liquifies, then made to flow into one or more mold cavities, where it is held at an elevated temperature until cross-linking is completed.

**IONOMER**
A thermoplastic polymer that is 'ionically cross-linked' by means of the transfer of electrons between atoms (ionic bonds).

**J**

### JUTE
A glossy fiber of the linden family used chiefly for making burlap and twine.

**K**

### KENAF
Long bast fibers from the family *Hibiscus cannabinus* native to India, suitable for papermaking.

### KEVLAR®
The brand name for a high-performance para-aramid fiber used as reinforcement and manufactured by Dupont.

### KRAFT PAPER
A strong and relatively cheap paper made chiefly from pine by using hot chemical solutions containing sodium sulfate (the kraft process) to remove the lignin and make the pulp.

### KYNOL®
The brand name for a range of phenolic fibers used for the production of high-performance materials including rovings, yarns, felts, fabrics and papers.

**L**

### LAMINATED GLASS
An impact-resistant material made by bonding sheets of transparent abrasion-resistant glass with a resilient plastic such as polyvinyl butyral (PVB) to produce a highly transparent safety glass for applications such as automotive windshields and protective goggles.

### LENO WEAVE
A fabric in which the warp yarns are paired and twisted.

### LENTICULAR PRINTING
Printed graphics that are laminated underneath a lenticular film to create a three-dimensional or flip image.

### LIGNIN
An amorphous polymer (phenylpropane) that comprises between 17 and 30 per cent of wood. It is used as a ceramic binder, as a dye leveler and dispersant, and for special molded products.

### LIQUID-CRYSTAL POLYMER
Polymers in which the molecular chains remain essentially in alignment even if the resin is molten at elevated temperatures or dissolved in a solvent.

### LYCRA®
A Dupont trademark for its spandex fiber, which has exceptional elasticity and durability.

**M**

### MDF
Medium-density fiberboard. MDF panels are made of fibrous raw material and generally have smoother surfaces and edges than particleboard.

### MAGNETORESTRICTIVE MATERIALS
Materials that exhibit a strain (undergo a deformation) when exposed to a magnetic field.

### MAKROFOL®
Bayer Materialscience's trademark for its polycarbonate film.

### MATRIX
The continuous or principal material in which another constituent is dispersed in a composite. For example, in a ceramic particle-reinforced aluminum composite, the aluminum is the matrix; in a glass fiber-reinforced epoxy composite, the epoxy is the matrix.

### MELAMINE
A thermoset plastic made from the resin of melamine and formaldehyde, which has excellent hardness, clarity and electrical properties.

### MEMBRANE
A thin sheet of natural or synthetic material that is microporous and acts as an efficient filter for very small molecules such as ions, water and other solvents, and gases.

### MICROFIBER
Fibers with extremely small diameters in the μ range. Microfibers as fine as 0.01 to 0.001 denier and not more than 0.3 denier are used to make man-made leathers with the properties and appearance of natural leather. Thicker microfibers of 0.1 denier are used to make a suede-like material with a fine nap.

**MICROSPHERES**
Hollow spheres in the µ size range (between 20 and 150 µ). They can be made of glass, silica, various polymers with high molecular weights (>5,000) or proteins (e.g. gelatin, albumen).

**MIL**
A unit of length equal to 0.001 inch.

**MODACRYLIC FIBERS**
Synthetic textile fibers that are long-chain polymers composed of between 35 and 85 per cent-by-weight acrylonitrile units.

**MODULUS**
A quantity that expresses the degree to which a material possesses a property such as elasticity.

**MONOFILAMENT**
A single, continuous length of fiber.

**MUNTZ METAL**
A specific form of brass, also known as alpha-beta brass, with a high zinc content.

# N

**NANO-COMPOSITES**
Composite structures in which the reinforcing particles are less than 100 nm in size.

**NEEDLE-PUNCHED**
A nonwoven textile made by passing barbed needles through a fiber web to entangle the fibers.

**NEOPRENE**
A synthetic elastomer (chemical name: polychloroprene) available as a solid, a latex or a flexible foam. Uses include rubber products and adhesive cements, seat cushions and carpet backing.

**NEXTEL®**
The brand name for a family of ceramic fibers used for composite reinforcement and the production of tapes, sleevings and textiles that can withstand elevated temperatures.

**NITRILE RUBBER (NBR)**
A class of rubber-like co-polymers of acrylo-nitrile with butadiene. They have high resistance to solvents, oils, greases, heat and abrasion.

**NOMEX®**
The trade name for Dupont's heat-resistant aramid fiber.

**NONWOVEN FABRIC**
A planar textile structure produced by loosely compressing together fibers, yarns, rovings and the like by mechanical, chemical, thermal or solvent means, or combinations of these.

# P

**PHENOLIC**
A theromosetting resin made by the polymerization of a phenol with an aldehyde. Phenolics are in the same family as polyesters, vinyl esters and epoxy resins.

**PHYSICAL VAPOR DEPOSITION (PVD)**
A type of vacuum-deposition process in which a material is vaporized in a vacuum chamber, transported atom-by-atom across the chamber to the substrate, and condensed into a film at the substrate's surface.

**PHOTOCATALYSIS**
A chemical reaction caused by impingement of light onto a material surface.

**PHOTOLUMINESCENCE**
The emission of visible or invisible radiation resulting from the absorption of energy in the form of photons from visible or invisible light.

**PHOTOPOLYMER**
A polymer or plastic that undergoes a change on exposure to light.

**PIEZOELECTRIC MATERIAL**
A piezoelectric material (e.g. quartz) converts mechanical stress into electrical energy and, conversely, expands in one direction and contracts in another when subjected to an electric field. Piezoelectric thin- and thick-film polymers and ceramics are used as transducers and sensors, as well as for microphones and earphones.

**PLASTICIZER**
A material (usually an organic compound) added to a polymer both to facilitate its processing and to increase the flexibility and toughness of the final product.

**POLYAMIDE (PA)**
The family of polymer resins that includes nylon. These polymers are strong, elastic and durable and are typically used in fiber form in apparel and upholstery and for garden implements in molded form.

**POLYANILINE**
A polymer that has electrical and heat-conducting properties. Typically this polymer is used in corrosion-inhibiting paint and as a packaging material for electronic products.

**POLYCARBONATE (PC)**
An engineering polymer resin that is a linear polyester of carbonic acid. Polycarbonate is a transparent, nontoxic, non-corrosive, heat-resistant, high-impact plastic and is commonly used as visors, non-breakable windows and household appliances.

**POLYETHYLENE**
A polymer resin from ethylene that is used in manufacturing trash bags, milk jugs, shampoo bottles and water coolers.

**POLYMERIZATION**
A chemical reaction, usually carried out with a catalyst, heat or light, in which a large number of relatively simple molecules (monomers) combine to form a chain-like macromolecule.

**POLYMETHYL METHACRYLATE (PMMA)**
A thermoplastic resin from the acrylic family of polymers that has good resistance to inorganic acids and alkalis but is attacked by a wide range of organic solvents. Typically used as a rigid sheet with a high degree of transparency as a tough alternative to glass.

## POLYPROPYLENE

A thermoplastic resin polymerized from propylene that is similar to polyethylene but significantly stiffer. It has excellent chemical resistance, is strong and has the lowest density of the plastics utilized for packaging. It has a high melting point and is used in the production of synthetic fibers, automotive parts, luggage and safety helmets.

## POLYURETHANE

A family of thermoplastic resins formed from the combination of a polyol with isocyanate. These resins typically have excellent abrasion and solvent resistance and a wide range of applications that include adhesives, paints, cabling, hosing and molded parts for automotive exteriors.

## POLYVINYL CHLORIDE (PVC)

A thermoplastic resin that is formed by the polymerization of vinyl chloride. Resins of polyvinyl chloride are hard, but a flexible, elastic plastic can be made with the addition of plasticizers. This plastic has found extensive use as an electrical insulator for wires and cables. Cloth and paper can be coated with it to produce fabrics that may be used for upholstery materials and raincoats.

## POST-CONSUMER RECYCLED CONTENT

Materials that are recovered from used and discarded consumer products and then recycled into new materials or products.

## POST-INDUSTRIAL PRODUCTION/PRE-CONSUMER CONTENT

Materials or by-products that have not reached a business entity or consumer for an intended use, including industrial scrap, overstock or obsolete inventories, and items generated from and reused within the original manufacturing process.

## PULTRUSION

A continuous process for manufacturing composites which have a constant cross-sectional shape. It consists of pulling a fiber-reinforcing material through a resin-impregnation bath and then through a shaping die, where the resin is subsequently cured.

# R
■

## RADIO-FREQUENCY INTERFERENCE (RFI)

Undesirable interference, particularly in electronic devices and equipment, from radiation from sources such as other electronic devices or electric motors. Most such problems are in the radio-frequency range (a frequency at which electromagnetic radiation of energy is useful for communication: approximately 10 KHz to 100 GHz.

## RAMIE

The flax-like cellulose fiber used in making fabrics and cordage, from the stem of the ramie plant, an Asian perennial herb.

## RAPID PROTOTYPING

Processes for quickly fabricating complex-shaped three-dimensional objects directly from CAD (computer-aided design) models. A number of such methods have been developed based on a concept called layered manufacturing or solid freeform fabrication (SFF), in which 3-D computer models of an object are decomposed into thin cross-sectional layers, followed by physically forming the layers and stacking them up to create the object.

## RASCHEL KNIT

A warp-knitted fabric in which the resulting knit resembles hand-crocheted fabrics, lace and netting.

## REFRACTORY (BRICK)

A highly heat-resistant and non-conductive material made of quartzite, high-silica clay or metallic ores (e.g. chromite, magnesite, zirconia) used for applications such as furnaces that operate at temperatures above 2,912∞F/1,600∞C.

## RESIN

A polymer produced by a chemical reaction between two or more substances, which does not contain any additives such as plasticizers, stabilizers or fillers.

## RETICULATED FOAM

Foams characterized by a three-dimensional skeletal structure with few or no membranes between strands; they may be polymer, ceramic, glass or metal. Reticulated foams are generally used as filters, acoustical panels or insulation.

## RETRO-REFLECTIVITY

Night-time visibility of, for example, pavement markings and highway and street signs when they reflect vehicle headlights.

## ROVING

A bundle of continuous filaments in the form of an untwisted strand or twisted yarn.

## RUBBER

A natural or synthetic high molecular-weight polymer with unique elongation properties and elastic recovery after vulcanization with sulfur or other cross-linking agents, which in effect converts the polymer from a thermoplastic to a thermoset.

## S

### SCRIM
A durable, loosely woven fabric, usually cotton or linen, used for drapery or upholstery lining or in industry. Also, a transparent fabric used to create special effects of light or of atmosphere.

### SHAPE-MEMORY POLYMER
A polymer that, following deformation, will return to its original shape when heated or cooled beyond a certain temperature.

### SHAPE-MEMORY METAL
A metal that, following deformation, will return to its original shape when heated or cooled beyond a certain temperature. Metal alloys that exhibit this phenomenon include nickel titanium and nickel iron. They find application in auto-off switches in electric kettles.

### SILICONE
Any of a large group of polymers called siloxanes that are based on a chemical structure consisting of alternate silicon and oxygen atoms with various organic chemical groups attached to the silicon atoms. They are liquids, semi-solids or solids and have a wide variety of applications including as adhesives, textile finishes, lubricants and coatings.

### SISAL
A strong bast fiber that originates from the leaves of the agave plant, which is found in the West Indies, Central America and Africa. Typical end-uses include cordage and twine.

### SLUBBY
A textile containing yarn that has soft, thick nubs which are either imperfections or purposely created for a desired effect.

### SOUND-ABSORBING
A material that dissipates the propagation of sounds by retaining vibrations in the range heard by the human ear (20 to 20,000 Hz) without reflecting or transmitting them. The propagation of sound can be dissipated by converting sound energy into thermal energy.

### SPALLING
The chipping, flaking, fragmentation or separation of a surface or coating.

### SPECTRA®
The trade name for Honeywell's oriented-polyethylene fiber. This high-tenacity fiber is typically used in ballistic armor.

### SPIN-BONDING
A process that forms nonwoven fabrics by adhering the fibers immediately after they are extruded from spinnerets.

### STITCH-BONDED
A textile in which fine lengthwise yarns in a warp knit are chain-stitched to interlock the fiber web-base structure or inlaid yarns.

### STRUCTURAL-COMPOSITE LUMBER (SCL)
Joists and other lumber produced using an oriented network of lumber strands laminated together with a waterproof adhesive to form a single, solid form.

### SUPERALLOYS
A family of metal alloys based on nickel cobalt or iron that can operate at elevated temperatures and in aggressive environments. These alloys were developed for use in the hot sections of gas-turbine engines.

## T

### TEFLON®
The trade name for Dupont's ethylene tetrafluoroethylene (ETFE) resin.

### TEMPERING
To harden or strengthen metal or glass by heating and cooling.

### TEMPERED GLASS
In the tempering process a high permanent stress is induced in glass by rapidly cooling the surfaces slightly below the softening-point temperature. This creates a high degree of compression at the surfaces with the tensile forces confined to the interior. Tempering greatly increases the strength of the glass and thus its impact resistance. One of the uses of tempered glass is for automotive windshields.

### TENACITY
The strength per unit weight of a fiber or filament, expressed as gram-force per denier (Gf/D) or Newton/tex (N/tex). It is measured as the rupture load or breaking force (gram-force or Newton) divided by the fiber's linear density (denier or tex).

### THERMAL-ARC SPRAYING
A coating technology in which two oppositely charged metal wires are fed through a gun, creating an electric arc when they touch that causes the metal to melt. A highly compressed gas fed through an orifice atomizes the molten metal and propels the droplets, spraying a thin layer of the metal onto the surface to be coated. The desired coating thickness is obtained by spraying multiple layers. This process does not produce undesirable combustion by-products.

### THERMAL SHOCK
The tendency of materials such as ceramics and glasses to partially or completely fracture, usually as a result of sudden or rapid cooling.

### THERMOFORMING
A thermoplastic sheet is formed into a three-dimensional shape after heating it to the point at which it becomes soft and flowable; then differential pressure is applied to make the sheet conform to the shape of a mold or die.

### THERMOFUSING (THERMAL WELDING)
Joining two or more pieces of material by applying heat.

### THERMOPLASTIC POLYMERS
Polymers that can be softened repeatedly by an increase in temperature and hardened by a decrease in temperature. When heated, thermoplastic polymers undergo a largely physical rather than a chemical change, and in the softened stage they can be shaped by molding or extrusion. Examples include polyethylene, polyvinyl chloride and polystyrene.

### THERMOSET POLYMERS
High molecular-weight polymers that can be cured, set or hardened by heat or radiation into a permanent shape. They form cross-links on heating and become permanently rigid, do not soften when reheated, but can decompose at high temperatures. Examples include polyesters and silicones.

### TOW
An untwisted bundle of continuous filaments, usually referring to man-made fibers, particularly carbon and graphite, but also to glass and aramid fibers (e.g. Kevlar®).

**TREVIRA®**
The trade name for Trevira's polyester fiber. Durable and inexpensive, it is used in apparel and upholstery.

**TRICOT KNIT**
A warp-knit fabric formed by inter-looping adjacent parallel yarns.

**V**

**VELLUM**
A type of cellulose-based paper that exhibits a rough and absorbent surface.

**VISCOELASTIC**
A material whose response to a deforming load combines both viscous and elastic qualities.

**VOLATILE ORGANIC COMPOUND (VOC)**
Any hydrocarbon (an organic compound consisting of carbon and hydrogen) except methane and ethane, with vapor pressure equal to or greater than 0.1 mm mercury. The VOC content refers to the amount of these constituents that will evaporate at their use temperature.

**VULCANIZATION**
A physical change in a rubber, such as polyisoprene, that improves its strength and resiliency and reduces its stickiness, resulting from a chemical reaction with sulfur or other suitable additives that cross-link the polymer chains.

**W**

**WARP**
In a woven fabric, the threads that run lengthwise and are crossed at right angles to the weft.

**WEFT**
The horizontal threads interlaced through the warp in a woven fabric. Also called the woof.

**WYZENBEEK TEST**
A precision test for determining the wear resistance of fabrics. In this test, a sample, which is held under a measured tension and a measured pressure, is subjected to the abrasive material clamped onto an oscillating drum operating at the rate of 90 double oscillations per minute.

**INFORMATION**
TRADE SHOWS + PROFESSIONAL PUBLICATIONS

## SHOWS

### COMPOSITES

**Composites 2005**
www.acmashow.org

**JEC Composites Show**
www.jecshow.com

### MATERIALS SCIENCE

**Materials Research Society Fall and Spring Meetings**
www.mrs.org

**Material Science and Technology**
www.matscitech.org

### PACKAGING

**PackExpo**
www.PackExpo.com

### TEXTILES

**Avantex**
www.avantex.de

**IFAI Expo (Industrial Fabrics Association International)**
www.ifaiexpo.info

**Techtextil**
www.techtextil.de

## PUBLICATIONS

### CERAMICS

*Advanced Ceramics Report*
www.intnews.com

*American Ceramic Society Bulletin*
www.ceramicbulletin.org

### COMPOSITES

*Advanced Composites Bulletin*
www.intnews.com

*CM Magazine (Composites Manufacturing)*
www.cmmagazine.org

*Composites Technology*
www.compositesworld.com

*JEC Composites Magazine*
www.jeccomposites.com

### GLASS

*Glass Magazine*
www.glassmagazine.net

*US Glass Magazine*
www.usglassmag.com

### METALS

*Advanced Materials and Processes*
www.asminternational.org

*Materials World*
www.iom3.org/materialsworld

*MRS Bulletin*
www.mrs.org/publications/bulletin/

## PLASTICS

*Injection Molding Magazine*
www.immnet.com

*Modern Plastics*
www.modplas.com

*Plastics Engineering*
www.4spe.org

*Plastics News*
www.plasticsnews.com

*Plastics Technology*
www.plasticstechnology.com

### TEXTILES/FIBERS

*Fabric Architecture*
www.aiasnatl.org

*GFR Engineering Solutions*
www.gfrmagazine.info

*Industrial Fabric Products Review*
www.ifai.com

*International Fiber Journal*
www.fiberjournal.com

*Nonwovens Industry*
www.nonwovens-industry.com

*Technical Textiles International*
www.technical-textiles.net

### GENERAL SCIENTIFIC INTEREST

*Nature*
www.nature.com

*New Scientist*
www.newscientist.com

*Scientific American*
www.sciam.com

## PACKAGING + MEDICAL MANUFACTURING

*Cosmetic Packaging and Design*
www.cosmeticpackaginganddesign.com

*Medical Product Manufacturing News*
www.devicelink.com/mpmn

*Pharmaceutical and Medical Packaging News*
www.pmpnews.com

### SUSTAINABILITY

*Environmental Building News*
www.buildinggreen.com

### TECHNOLOGY

*Air Force Research Laboratory Technology Horizons*
www.afrlhorizons.com

*MIT Technology Review*
www.techreview.com

*Nasa Tech Briefs*
www.techbriefs.com

I WANT TO THANK MY PUBLISHER, JAMIE CAMPLIN, FOR CONSISTENTLY ENCOURAGING ME TO PRODUCE THIS BOOK. ON THE SUBJECT OF PUBLISHERS, I WAS LUCKY TO HAVE EXPERIENCED ANOTHER EXEMPLARY ONE WHO CHAMPIONED THE CAUSE OF MATERIALS: THE LATE PAUL GOTTLIEB OF HARRY N. ABRAMS, INC., WHO PUBLISHED MONDO MATERIALIS IN 1980. I WANT TO EXPRESS MY APPRECIATION FOR TWO COLLABORATORS WHO CONTRIBUTED TO MATERIAL CONNEXION AS WE STARTED OPERATIONS IN 1997: DR. RENÉE FORD AND LÌAT MARGOLIES. THE EXPERIENCE OF COLLABORATING WITH MY CO-AUTHOR, ANDREW DENT, AND MY EDITOR, ANITA MORYADAS, HAS RESULTED IN A TRULY FLAWLESS WORKING RELATIONSHIP; HOPEFULLY THE RESULTS ARE REFLECTED HEREIN FOR THE READER'S ENJOYMENT.

G.M.B.
NEW YORK
2004

## PHOTOGRAPHIC CREDITS

The materials illustrated in this book were photographed by John Michael Ekeblad and Eugene Gologursky.

The authors would like to acknowledge the following for permission to reproduce photographs of objects, interiors, installations and buildings: **p. 9** (Wire Basket): Abbeville Press; **p. 10** (Flexform Special Finish): Courtesy of Philips Design; (*Untitled*): Lisa Mangino; **p. 11** (I Feltri): Courtesy of Gaetano Pesce for Cassina; **p. 13** (*Waltz of the Polypeptides*): James Brown; (*Red Tree*): Alistair Overbruck; (Gel Mask): Courtesy of Gaetano Pesce; **p. 16** (Cardboard Pod): Timothy Hursely; **p. 17** (Off the Street, from the Beach): Bobby Hansson; (Crab-claw Necklace): Doug Van De Zande; (Geisha Doll) Lisa Mangino; (Mountain Goat): Seiji Kakizaki; **p. 18** (Little Hall): Barry Hakins; **p. 19** (Airtecture + Airquarium): Festo AG; (Mobile Dwelling Unit): Courtesy of Lot-Ek; **p. 20** (Trousers): Lisa Mangino; **p. 21** (Scrapyard): Mr Takao Inoue; **p. 24** (Dimension Polyant): Ken Madsen; **p. 25** (Talon): Talon; **pp. 32-3**: John Michael Ekeblad; **p. 41** (Agrob Buchtal): Agrob Buchtal; (Royal Tichelaar Makkum): Courtesy of Royal Tichelaar Makkum; (Doug Fitch): Dean Powell; **pp. 50-51** (Maya Romanoff): Darrin Haddad; **p. 52** (FTL): Durstan Saylor; p. 53 (Andy Cao): Courtey of Andy Cao; **p. 55** (Omnidecor): Omnidecor Spa; **p. 59** (Oceanside Glass Tile): Courtesy of Oceanside Glass Tile; **p. 69** (Johns Manville): Courtesy of John Mansville; **pp. 76-7** (Moradelli): Courtesy of Moradelli; **pp. 78-9** (Cobra): Eugene Gologursky; (Mabeg [Design Afairs]): Eugene Gologursky; **p. 86** (Peer Inc.): Courtesy of Peer Inc.; **p. 92** (Carl Stahl): Carl Stahl; **p. 96** (Haver & Boecker): Courtesy of Haver & Boecker; **p. 99** (Panelite Laminates): Courtesy of Panelite LLC; **p. 101** (KME): KM Europa Metal AG; **p. 102** (Moz Designs Inc.): Phillip McCain; **p. 105** (Ball Chain Manufacturing Co. Inc.): Ball Chain Manufacturing Co. Inc.; **pp. 108-9** (Spinneybeck): Sandy Levy; (Abet Laminati): Abet Laminati; **pp. 110-11**

(Materalise; Mio Culture; Aveda; Reholz): Eugene Gologursky; **p. 128** (G.T. Design): Deanna Comellini for GT Design; **p. 129** (Ruckstuhl): Courtesy of Ruckstuhl; **pp. 136-7** (Topakustik): Alessandro Paderni – Studio Eye; **p. 146** (Effe Marmi Spa): Effepimarni SRL; **p. 148** (Creation Bauman): creation bauman; **p. 154** (Morning): Albi Serfaty; **p. 155** (Aqua Gallery): Courtesy of Aqua Creations; **pp. 156-7** (Bouroullec Bros): Ronan Bouroullec; (Flex Products): Eugene Gologursky; (Geltec): Supplied by Geltec Co., Ltd.; (Nike): Eugene Gologursky; **p. 167** (Fuseproject): Marcus Hanschen; **p. 172** (Tizip): ORTLIEB Sportartikel GmbH; **p. 177** (Trespa): Trespa International; **p. 178** (Ecco): Ken Stalski; (Boffi [Marcel Wanders]): Duilio Bitetto for Boffi; **p. 179** (Trozdem): Courtesy of Ingo Maurer; **p. 183** (Honeycomb Cushioning): Surpacor, Inc.; **p. 197** (Cachet Chair): Steelcase Inc.; (Paola Lenti): Courtesy of Paola Lenti; **p. 198** (Luminex Shirt): Courtesy of Luminex; **pp. 200-201** (Dakota Jackson): Dakota Jackson, Marissa Brown; **p. 205** (Built NY): Courtesy of Built NY; **p. 213** (Nissan): Deborah Forester; **p. 220** (Tord Boontje): Courtesy of Artecnica; **p. 229** (em2): Courtesy of Philips Design; **pp. 242-3** (Moroso [Ron Arad]): Moroso; (Atta, Resin Bar): Tori Butt; (Atta, Resin Stair): Peter Aaron/Esto; **p. 244** (Field Turf): Field Turf; **p. 245** (Snowbird): Eugene Gologursky; **p. 249** (Designtex): Designtex.

**INFORMATION**
INDEX